LABORATORY MANUAL

to accompany

Refrigeration and Air Conditioning, Third Edition

Greg Jourdan
Air-Conditioning and Refrigeration Institute
4301 North Fairfax Drive, Suite 425
Arlington, VA 22203

Prentice Hall
Upper Saddle River, New Jersey *Columbus, Ohio*

Editor: Charles E. Stewart, Jr.
Developmental Editor: Carol Hinklin Robison
Production Editor: Stephen C. Robb
Design Coordinator: Julia Zonneveld Van Hook
Cover Designer: Brian Deep
Production Manager: Pamela D. Bennett
Marketing Manager: Danny Hoyt

This book was was printed and bound by Courier/Kendallville, Inc.
The cover was printed by Phoenix Color Corp.

Printed in the United States of America

10 9 8 7 6 5 4 3 2 1

ISBN: 0-13-646407-6

Prentice-Hall International (UK) Limited, London
Prentice-Hall of Australia Pty. Limited, Sydney
Prentice-Hall of Canada, Inc., Toronto
Prentice-Hall Hispanoamericana, S. A., Mexico
Prentice-Hall of India Private Limited, New Delhi
Prentice-Hall of Japan, Inc., Tokyo
Simon & Schuster Asia Pte. Ltd., Singapore
Editora Prentice-Hall do Brasil, Ltda., Rio de Janeiro

GENERAL INSTRUCTION

LABORATORY TASKS

The tasks in this Manual are sequenced to follow the chapter arrangement in the text book "Refrigeration and Air Conditioning by ARI" as closely as possible. The arrangement is such, however, that some flexibility in assignment is possible to allow closer coordination with student progress. The tasks in this revision have been modified to reflect required competencies of the HVACR Technician. Be sure to never intentionally vent refrigerants to the atmosphere and always recover and recycle refrigerants. Several new competencies have been added to reflect modern practices, including ice machines, hydronic heating, and DDC electronic control systems.

INSTRUCTIONS TO THE STUDENT

Laboratory Work Assignments

The instructor will give the laboratory assignments. This will be done either in a group or in working pairs, depending upon the subject involved. Each group will be assigned to a work station and will be expected to draw the tools and equipment necessary to perform the task involved. Each group will also be expected to clean up the work station after the task is completed.

1. Review the study material assigned.

2. Obtain the necessary equipment to perform the task.

3. Follow the procedure as outlined in the assignment sheet. Any question should be referred to the instructor before proceeding with the task.

4. Upon completion of the task, answer any questions that are given on the task sheet.

5. Clean up and put away all tools and equipment involved. Leave the work station clean and orderly for the next group.

OBSERVE SAFETY RULES AT ALL TIMES

1. Always wear safety goggles when working around equipment.
2. Only apply heat to refrigerant cylinders by applying warm water to the cylinder. Never use an open flame on the cylinder.
3. Be aware of loose or un-insulated electrical connections. Avoid touching belts, flywheels, blowers or fans or other moving parts.
4. Avoid the use of tools and equipment that are not in safe operational condition.
5. Refrigerant cylinders should never be filled more than 80% of their capacity. If this quantity is exceeded, the cylinder could burst from hydrostatic pressure. Also, keep cylinders away from high temperatures which could also raise the refrigerant pressure within the cylinder and cause over-pressurization and bursting.
6. When checking for leaks, the area must be well ventilated. Never use a Halide torch in a confined area such as a walk-in cooler. Fumes from a Halide torch are poisonous and can accumulate to a deadly quantity.
7. Before charging a unit, always make sure the supply cylinder has the proper refrigerant as required by the unit.
8. Liquid refrigerant on the skin may freeze the skin surface and cause frost burn. If this should occur, wash the area with water--treat the area the same as frost bite.
9. Refrigerant in the eyes can cause blindness. Flush immediately with water to remove the refrigerant. Hospitalization is a must and should be done immediately.
10. Be very careful of the oil from burnt motor compressor assemblies. Test with litmus paper or acid test kit before handling. Face shield, rubber apron and rubber gloves are required to prevent severe burns. If severe conditions exist, it may be necessary to use SCBA, Self Contained Breathing Apparatus.
11. Never intentionally vent refrigerant to the atmosphere. It is illegal and dangerous. It will cause oxygen deprivation and asphyxiation.
12. Always recover and recycle refrigerant when you have to replace or repair refrigeration components. If the refrigerant is severely contaminated, send it out to be reclaimed to ARI standards.
13. Chemical should be handled according to the manufacturer's specifications.
14. Technicians must use every safety precaution possible when working with pressures, electrical energy, heat, cold, rotating machinery, chemicals, and moving heavy objects.
15. Electrical energy is present while troubleshooting energized electrical circuits. Be careful at all times. Practice lock-out and tag-out safety procedures, if possible, when working on electrical components. Lock out panels or disconnect boxes and keep the only key in your possession or with your immediate supervisor.
16. When moving heavy equipment, use correct techniques, appropriate tools and equipment, and wear a back brace belt.
17. Never solder tubing lines that are sealed. Open service valves or Schroeder ports before soldering is attempted.
18. Technicians should use only properly grounded power tools connected to properly grounded circuits.

TABLE OF CONTENTS

SECTION: REFRIGERATION

SECTION: HEAT PUMPS

Laboratory Worksheet - # R1
Student____________________
Instructor ____________________
Date __________ Grade________

MATTER AND HEAT BEHAVIOR

EXPECTED COMPETENCY

To become acquainted with the temperature relationship to the state of a material

STUDY MATERIAL

Chapters R3 and R4

EQUIPMENT

1. A metal container of ice in which a dial type thermometer has been frozen, one Fahrenheit and one Celsius
2. Electric hot plate heat source
3. Scale
4. Safety goggles and gloves

SAFETY TIP

During this lab, be careful not to touch the boiling water, the metal container, or the hot plate. Remember to wear your safety goggles and gloves at all times.

PROCEDURE

1. Wearing the safety goggles, record the weight of the container and ice.
2. Place the container on the hot plate with the hot plate plugged into the proper electrical supply.
3. Record the following temperatures as they occur. As the ice melts, be sure the thermometer does not touch the side or bottom of the container.
   a. Temperatures of ice when placed on heat source.
      °F__________
      °C__________
   b. Temperature at which ice starts to melt.
      °F__________
      °C__________
   c. Temperature at which the last of the ice melts.
      °F__________
      °C__________
   d. Temperature at which boiling starts.
      °F __________
      °C__________

4. Pour out the last of the water in the container and weigh the empty container.
   Using the proper formulae, from the text book, calculate the amount of heat necessary to:
   (1) Melt the ice.
   (2) Bring the water to a boil.
   (3) Boil off the entire quantity of water.

DO NOT LEAVE THE CONTAINER ON THE HOT PLATE WITHOUT WATER IN THE CONTAINER.

QUESTIONS

1. The actual weight of the water was _______pounds.

2. The temperature of the ice from the freezer was _____°F and _____°C.

3. Raising the temperature of the ice from the freezer to the melting point of ______°F required _______B.T.U.'s of heat.

4. To melt the quantity of ice required ______B.T.U.'s of heat.

5. To raise the temperature of the water from the melting point to the boiling point required _______ B.T.U.'s of_________heat.

6. To boil off the water required _______B.T.U.'s of________ heat.

7. At the melting point of the ice, the temperature of the ice was _____°F and _____°C.

8. At the boiling point of the water, the temperature was _____°F and _____°C.

FOR THE FOLLOWING QUESTIONS, SHOW WORK IN SPACE BELOW.

9. Using the conversion formula, solve for the Celsius temperature using the Fahrenheit temperature at the melting point and at the boiling point.
   Melting_______ °C Boiling ________ °C

10. Using the conversion formula, solve for the Fahrenheit temperature using the Celsius temperature at the melting point and at the boiling point.
    Melting _______°F Boiling ________ °F

Laboratory Worksheet - # R2

Student______________________________

Instructor ____________________________

Date ______________ Grade__________

HEAT TRANSFER-CONDUCTION

EXPECTED COMPETENCY

To demonstrate the ability of material to transfer heat by conduction

STUDY MATERIAL

Chapter R4

EQUIPMENT

1. Safety goggles
2. Rods of various materials
   a. Copper
   b. Steel
   c. Glass
3. Bunsen burner (or equivalent) heat source
4. Stop watch

SAFETY TIP

Be cautious when working around the heat source and various materials to prevent burning yourself.

PROCEDURE

1. With the rod held at one end, start the stop watch and insert the opposite end of the rod in the burner flame. As soon as the heat is felt at the end of the rod, remove it from the flame and put it on a heat resistant surface to prevent burns.
2. Record the time needed for the heat to be felt at the held end of the rod.
   a. Time for copper _______ sec.
   b. Time for steel _______ sec.
   c. Time for glass _______ sec.

QUESTIONS

1. Which material conducted the heat in the least time?

2. Why does the glass rod require the greatest amount of time?

3. What are the applications for copper, steel, and glass materials in the HVACR industries?

Laboratory Worksheet - # R3
Student___________________________
Instructor ___________________________
Date ____________ Grade__________

CONVECTION CURRENTS

EXPECTED COMPETENCY
To demonstrate the transfer of heat by convection currents

STUDY MATERIAL
Chapter R4

EQUIPMENT
1. Clear glass container of Pyrex or heat resistant type
2. Small quantity of paper punchings
3. Electric hot plate

SAFETY TIP
Be cautious when working around the heat plate source to prevent burning yourself. Remember to wear safety goggles.

PROCEDURE
1. Fill the glass container approximately 75% with water.
2. Place halfway on electric hot plate. Plate should heat only 50% of thc bottom of the container.
3. Place a few paper punchings in the water and observe the action.

QUESTIONS
1. What causes the action of the heat added to the water in the section of the container over the hot plate?

2. Why does this action take place?

Laboratory Worksheet - # R4
Student____________________________
Instructor __________________________
Date ____________ Grade__________

RADIANT HEAT TRANSFER

EXPECTED COMPETENCY
To demonstrate the effect of heat radiation regarding travel distance and color absorption rate

STUDY MATERIAL
Chapter R4

EQUIPMENT
1. Radiant heat source: Electric radiant heater
2. Pieces of colored cloth
    a. Black
    b. White
    c. Red
    d. Green
3. Stop watch

SAFETY TIP
Be cautious when working around the heat source to prevent burning yourself. Remember to wear safety goggles.

PROCEDURE
1. Set the heater on the table, measure and mark off distances of two, four and six feet from the front of the heater.
2. Connect the heater to an appropriate power supply.
3. With a cloth dropped over your hand, record the time it takes for the heat to be felt from the distance of six feet, four feet and two feet. Change colors each time to allow the cloth to cool to room temperature.

QUESTIONS
1. Which color cloth absorbs the heat fastest?

2. At what distance is the heat absorbed fastest?

Laboratory Worksheet - # R5
Student______________________
Instructor ______________________
Date ____________ Grade_________

FLUIDS AND PRESSURE

EXPECTED COMPETENCY
To demonstrate the effect of temperature on volume

STUDY MATERIAL
Chapter R5

EQUIPMENT
1. Tall drinking glass
2. Aluminum pie pan
3. Hot water source
4. Cold water source
5. Safety goggles

SAFETY TIP
Be cautious when handling the glass so as not to break it near your work station.

PROCEDURE
1. Wearing the safety goggles, place drinking glass upside down in the aluminum pie pan.

2. Place the glass and pan under a slow hot water faucet until glass reaches water temperature. Air will stop leaving the rim of the glass.

3. Change the water source from hot to cold. Leave under cold water until water reaches maximum height in the glass.

QUESTIONS
1. Why did the air leave the glass when the glass was subjected to the hot water?

2. Why did the water build in the glass when under the cold water?

3. What law does this action, due to temperature change, demonstrate?

Laboratory Worksheet - # R6
Student___________________________
Instructor _________________________
Date _____________ Grade__________

MEASURING AIR PROPERTIES

EXPECTED COMPETENCY
To demonstrate the building and action of a manometer

STUDY MATERIAL
Chapter H40

EQUIPMENT
1. 12" ruler
2. 4' of clear plastic hose - %" diameter
3. Quantity of water
4. Air distribution system (heating or cooling duct system)
5. Clear plastic tape

SAFETY TIP
Be cautious when working around the heat source and various materials to prevent burning yourself.

PROCEDURE
1. Form the plastic tube into a "U" shape similar to the shape of the ruler.
2. Holding the ruler on end with the 1" mark on top, form the plastic tube to the shape of the ruler. The short open end and the long connection tube should be at the top of the ruler. See example in text.
3. Tape the plastic tubing to the ruler at the 1", 4", 8" and 12" marks.
4. Fill the tube with water so that the water levels out the 6" mark.
5. Using this water manometer, measure the pressure in the duct of the air distribution system.

QUESTIONS
1. If the duct pressure forces the water level 1" below the 6" mark, what is the air pressure in the duct?

2. When connected to the return air system, the water level is 3/4" above the 6" mark. What is the pressure in the return air duct?

3. Is the pressure in the return air duct a positive or negative pressure?

Laboratory Worksheet - # R7
Student________________________
Instructor ______________________
Date ____________ Grade_________

PRESSURE/TEMPERATURE RELATIONSHIP

EXPECTED COMPETENCY:
To demonstrate the effect of pressure on boiling point

STUDY MATERIAL
Chapter R5

EQUIPMENT
1. Bell jar
2. Vacuum plate
3. Small sauce pan
4. Immersion type thermometer - dial type
5. Hot water from sink faucet
6. Vacuum pump
7. Set of vacuum gauges
8. Small quantity of petroleum jelly

SAFETY TIP
Keep safety goggles on at all times, even when around low pressures.

PROCEDURE
1. Connect vacuum plate to vacuum pump and vacuum gauge. Check by placing plug in hole of vacuum plate and reading vacuum gauge. It should begin to pull a vacuum to at least 10". If it appears to be operating properly, stop the pump and remove the plug.
2. Place the pan on the vacuum plate.
3. Place the dial type thermometer in the pan with dial on the edge so that the dial can be seen through the bell jar.
4. Pour hot tap water into the pan to cover the stem of the thermometer .
5. Cover the edge of the bell jar with petroleum jelly for seal and place over the pan setup.
6. Operate the vacuum, watching the vacuum gauge and thermometer, until water starts to boil.
7. Record both the vacuum and temperature at which boiling took place.

QUESTIONS
1. Why did the water boil at the lower temperature?

2. The boiling point of water changes in the same direction as the pressure on the water. True or False?

3. What is the purpose of the pressure cap on the radiator of a car and how is the pressure/temperature relationship affected in comparison to a refrigeration system?

Laboratory Worksheet - # R8
Student__________________________
Instructor _________________________
Date ____________ Grade_________

REFRIGERATION PIPING MATERIALS IDENTIFICATION

EXPECTED COMPETENCY:
To become acquainted with the various pipes and fittings used in the refrigeration system

STUDY MATERIAL
Chapter R16

EQUIPMENT
1. Various size pieces of iron pipe, both black and galvanized
2. Various size pieces of copper tubing, both soft-drawn and hard-drawn
3. Samples of copper fittings - sweat type
4. Samples of copper fittings - flare type
5. Samples of iron pipe fittings
6. Samples of compression fittings
7. Samples of diaphragm fittings

SAFETY TIP
Be cautious when working around freshly cut copper and steel pipe. Sharp edges may protrude from the end and cut your hands.

PROCEDURE
List and become familiar with each type of fitting.

QUESTIONS
1. What is the dimension difference between copper pipe and iron pipe?

2. Does the refrigeration service person talk I.D. or O.D. when specifying copper tubing?

3. Does the plumber talk I.D. or O.D. when specifying iron pipe size?

4. What are the three classifications of copper tube?

5. Which two are used in refrigeration systems?

6. What is the meaning of "refrigeration" grade copper tubing?

7. Which type of copper may be bent?

8. How is soft copper different from hard drawn copper when delivered to the job site?

Laboratory Worksheet - # R9

Student__________________________

Instructor __________________________

Date _____________ Grade__________

FLARING COPPER TUBING

EXPECTED COMPETENCY:

To become acquainted with the procedures involved in the making of satisfactory flare connections

STUDY MATERIAL

Chapter R16

EQUIPMENT

1. Tubing - 1/4", 3/8", ½" and 5/8"
2. Flare nuts - 1/4", 3/8", ½" and 5/8"
3. Flaring tool set
4. Oil can

PROCEDURE

1. Using the 1/4" tube, if there is any indication of an inner "lip" as a result of cutting the copper tube, ream the tubing, holding the tube downward to prevent chips from falling into the tube.
2. Slide the 1/4" flare nut on the 1/4" tube with the threaded female part toward the end of the tube to be flared.
3. Place the 1/4" tube in the flaring tool so that it extends 1/3 as much as the depression in the flare block.
4. Place the flaring reamer above the opening in the tube and put a drop of oil on the reamer before turning it tight.
5. Make the flare with an oscillating motion, making sure not to put too much pressure at one time on the tube. Draw the flare. DO NOT FORCE IT.
6. After the reamer and flare have reached the block, remove the flare and slide the nut over the flare. The flare width should slide through the threads of the nut without excessive clearance.
7. Have the flare inspected by your instructor(s) so that they can point out any possible errors in size, cracking, reaming and thickness.
8. Follow the same procedure with the 3/8", ½" and 5/8" tubing, noting that the larger the tubing the greater the extension above the flare block compared to the 1/4" tube. Note also the larger size tubing will crack easier and requires more care in forcing the reamer into the tube.

QUESTIONS

1. Name two methods of cutting copper tubing.

2. What causes flares to split and how can this be avoided?

3. If a flare is too small, what should be done?

4. What happens if the tubing is tightened down too fast while cutting it with a tubing cutter?

5. Why should you apply refrigerant oil to the flare cone before making a flare?

6. What is the purpose of reaming the tube before flaring?

7. What actually seals the connection in a flare connection?

Laboratory Worksheet - # R10
Student________________________
Instructor ______________________
Date ____________ Grade_________

MAKING SWAGE CONNECTIONS

EXPECTED COMPETENCY:
To become acquainted with procedures involved in making swage joints

STUDY MATERIAL
Chapter R16

EQUIPMENT
1. Copper tubing - 1/4" , 3/8" and 5/8"
2. Swaging tool
3. Flaring block
4. Hammer

PROCEDURE
1. Place the 1/4" tube in the flare block so that it extends above the block slightly more than the diameter of the tubing.
2. Insert the 1/4" swaging tool in the tubing, hold the flare block and tubing in one hand, and tap the swaging tool with the hammer. MAKE SURE THE FLARE BLOCK IS HELD IN THE HAND - it must not be clamped in a vice or on anything solid.
3. Extreme caution must be observed to keep the swaging tool aligned with the tubing at all times and it must be tapped squarely with the hammer. If the swaging tool becomes slanted or out of line, it will make the swage joint lopsided and will make a poor joint. Check frequently as you are making the swage joint.
4. When the swage joint is completed, test it for proper fit by inserting another piece of tubing, of the same size, into it.
5. Have the swage joint inspected by the instructor(s) so that they can point out any possible errors in its construction.
6. When you have made an acceptable 1/4" swage joint, repeat the process with the 3/8" and ½" tube.

QUESTIONS
1. Name an important advantage of swage joints over other types of tube joints or connections.

2. Name a disadvantage of swage joints as compared with flare connections.

3. What happens when the swaging tool is struck too hard with the hammer?

Laboratory Worksheet - # R11
Student_______________________
Instructor _____________________
Date ___________ Grade_________

MAKING SOLDER CONNECTIONS

EXPECTED COMPETENCY:
To develop skills in making soldered connections.

STUDY MATERIAL
Chapter R16

EQUIPMENT
1. Copper tubing - 1/4" , 3/8" and ½"
2. Swaging tool
3. Flare block
4. Hammer
5. 50/50 solder (for joint examination)
6. 50/50 solder flux
7. Low temperature silver solder
8. Silver solder flux
9. Sand cloth and fitting brush
10. Oxygen/acetylene welding setup or turbo (acy/air) torch
11. Hacksaw with 32 teeth per inch blade

SAFETY TIP
Be cautious when working around the welding torch. Wear proper eye protection and gloves as required. Follow normal pressure settings on welding tanks within acceptable parameters.

PROCEDURE
1. Using two pieces of 1/4" tube, make a swage joint in one of the pieces.
2. Clean both the male and female ends of the tubing to be joined. When cleaning, hold the tube end downward to prevent particles from entering the tubing.
3. Apply flux to the male tube only, being careful not to get any flux inside the tube.
4. Join the two pieces of tubing together to prepare to weld.
5. Place the tube with the female end in a vise or other clamp in an upright position.
6. Heat the swage joint with the proper size and type of flare. When the flux starts to bubble, apply the solder to the joint on the side opposite the flame.
7. Make sure that, as the solder solidifies, the copper is not moved or joined. This will cause a joint that is weak and could leak. If this should happen, reheat the joint to the melting point of the solder and let cool without jarring.
8. Using the hacksaw, cut the joint at a 45° angle through the center of the joint. Take it to the instructor for inspection and approval.

9. Repeat steps 1 through 7 using the 3/8" and ½" tubing.
10. Repeat steps 1 through 8 using the 1/4" , 3/8" and ½" tube and the low temperature silver solder and flux.

QUESTIONS

1. What is the purpose of using the flux?

2. Neither steel wool nor any cloth should be used to clean tubing. Why?

3. What are the advantages of solder connections over flared connections?

4. What are the disadvantages?

5. What is the disadvantage of using 50/50 solder in refrigeration systems?

6. What is the advantage of using low-temperature silver solder over high-temperature?

7. What are the appropriate temperature ranges used when soldering the following metals?
   a. Copper to copper ______________ °F
   b. Copper to steel ______________ °F
   c. Aluminum to aluminum ______________ °F

8. What is the minimum temperature for brazing?

Laboratory Worksheet - # R12

Student__________________________

Instructor ________________________

Date ____________ Grade_________

TUBE BENDING

EXPECTED COMPETENCY:

To become acquainted with the proper methods of bending soft-drawn copper tubing used in refrigeration systems

STUDY MATERIAL

Chapter R16

EQUIPMENT

1. 1/4", 3/8", ½" and 5/8" copper tubing
2. Tube bender
3. Bending springs

PROCEDURE

1. Make a 360° bend with the 1/4" tube. What is the smallest diameter circle that can be made with 1/4" tube?
2. Make a 360° bend with 3/8" tube. What is the smallest diameter circle that can be made with 3/8" tube?
3. Using the correct size bending spring, form a 270° bend with 3/8" tubing. Remember to release the bending spring, bend the tubing beyond the angle desired and return the tube to the angle desired. This action reduces the outside dimension of the tube and frees the spring.
4. Using the correct size bending spring, form a 180° hairpin bend using V copper tubing.
5. Use the tubing bender to form 90° bends using 3/8", ½" and 5/8" tubing. Compare the radius of the bend with the bends using the bending spring and by hand.
6. Have the bending projects inspected by the instructor(s) so they can point out any possible faults in their formation.

QUESTIONS

1. In refrigeration work, when a piece of tubing is cut from a coil, how must you treat the end?
2. Explain why this is necessary.
3. Where can an outside bending spring be used?
4. How are bends made using hard-drawn tubing?

5. Describe procedures for bending soft copper tubing.

6. What does annealing mean when working with soft copper?

Laboratory Worksheet - # R13

Student ____________________________

Instructor __________________________

Date ____________ Grade__________

TUBING PROJECT

EXPECTED COMPETENCY:

To use the skills learned in projects R8 through R12

STUDY MATERIAL

Chapter R16

EQUIPMENT

1. Copper tubing 1/4", 3/8" and ½"
2. Solder tees 1/4" and 3/8"
3. Flare nut 1/4"
4. Swaging tool
5. Flaring tool
6. Hammer
7. Low temperature silver solder
8. Flux
9. Oxygen/acetylene soldering outfit or propane turbo-torch
10. Tube bender for 1/4 , 3/8" and ½" tube
11. Set of tube-bending springs
12. Vice grip pliers
13. Nitrogen tank

SAFETY

Be cautious when working around the heat source and various tubing materials to prevent burning yourself.

PROCEDURE

1. Using the copper tubing layout drawing, fabricate each piece of tubing to size and shape. Allow additional tube length for insertion in joints.
2. Instructor will demonstrate the method of joining two tubes in a third tube.
3. Pressure test, using nitrogen with soap solution.

QUESTIONS

1. How many leaks did you find?

2. Where were the leaks found?

3. If you had any leaks, identify what caused the leaks.

4. What was necessary to repair the leaks?

Laboratory Worksheet - # R14
Student__________________________
Instructor __________________________
Date _____________ Grade__________

COMPRESSION CYCLE COMPONENTS

EXPECTED COMPETENCY:
To acquaint the student with the various parts of the refrigeration system and their basic functions

STUDY MATERIAL
Chapter R7

EQUIPMENT
1. Window air conditioner with outer cabinet removed

SAFETY
Be cautious when working around the coil surfaces. Finned tubing tends to have sharp edges and will easily cut the skin during repairs and maintenance.

PROCEDURE
1. Name and explain the purpose of each component part.
2. Name the various forms or shapes of each manufactured component.
3. Name the ports required through which refrigerant flows to produce the desired refrigerating effect.
4. Identify the piping and direction of refrigerant flow through the complete refrigeration cycle.
5. Make sure correct air flow passes across each coil before starting the compressor.
6. Place a thermometer at the inlet and outlet of the evaporator and start the compressor.
7. If the unit is inside the conditioned space during this lab, cover the condenser coil approximately 50% with cardboard or paper to artificially raise the head pressure.
8. Let the unit run for about 15 minutes.
9. Notice whether the evaporator is sweating equally across the coil.
10. Record evaporator air temperatures. Entering air is ____°F Leaving air ____°F
11. Record the temperature difference across the condenser coil. ______°F

QUESTIONS
1. What type of compressor was used on the unit?

2. Explain the type of metering device.

3. What type of heat exchanger was used for the evaporator and the condenser?

4. Were there any other accessories included with the unit? (e.g.-accumulators, mufflers, filter/driers, limit controls, etc.)

5. What is the primary maintenance needed on room cooling units?

6. Draw the refrigeration cycle and the components below. Explain the operation of each component, direction of refrigerant flow, and pressures and temperatures in the system.

Laboratory Worksheet - # R15

Student____________________________

Instructor ___________________________

Date ______________ Grade___________

PRESSURE/TEMPERATURE RELATIONSHIP OF REFRIGERANTS

EXPECTED COMPETENCY:

To become acquainted with the commonly used refrigerants and their pressure/temperature relationship

STUDY MATERIAL

Chapter R13

EQUIPMENT

1. Drums of R-134a, R-401a, and R-22
2. Pocket pressure temperature charts
3. Pressure gauges with 1/4" flare couplings
4. Thermometer
5. Safety goggles

SAFETY

Be cautious when working around refrigerant cylinders to prevent leaking refrigerant to the atmosphere. Always wear your safety glasses and protective gloves when handling refrigerants.

PROCEDURE

1. Using the thermometer to determine the dry-bulb temperature of the room.
2. Wearing safety goggles, connect a gauge to each drum.
3. By reading the gauges and using the room temperature, determine from the pressure/temperature chart the type of refrigerant in each drum.
4. Place the drums in a refrigerator or walk-in cooler.
5. Observe the pressure change in each drum as the temperature of the drum drops.

QUESTIONS

1. What is the pressure in each drum at room temperature?

   a. R-134a ___________PSIG

   b. MP39 ___________ PSIG

   c. R-22 ___________ PSIG

2. What is the pressure in each drum at a lower temperature __________°F?
   a. R-134a__________ PSIG
   b. MP39 __________ PSIG
   c. R-22 __________ PSIG

3. Determine the amount of temperature drop for the test.
   a. Room temperature minus cooler temperature = ___________°F.

4. Determine the amount of pressure drop for each refrigerant between the two tests.

Refrigerant	High Pressure - Low Pressure = Difference
a. R-134a	_______PSIG - _______PSIG = _______PSIG
b. MP39	_______PSIG - _______PSIG = _______PSIG
c. R-22	_______PSIG - _______PSIG = _______PSIG

Laboratory Worksheet - # R16
Student_______________________
Instructor _____________________
Date ____________ Grade_________

MODERN REFRIGERANTS

EXPECTED COMPETENCY:
To demonstrate the ability to understand the properties and applications of the HFC and HCFC refrigerants. Also, to understand associated terminology, including: glide, fractionation, bubble point, and dew point.

STUDY MATERIAL
Chapter R13

EQUIPMENT
1. Safety goggles
2. A cylinder of R134a, 401a (MP39), 404a (HP62), R-22
3. 4 Refrigeration gauge manifolds
4. 4 Electronic thermometers
5. Pressure temperature (PT) chart for each refrigerant

SAFETY
Be cautious when working with the refrigerants to avoid releasing any gases into the atmosphere.

PROCEDURE
1. Install your gauges on each refrigerant cylinder.
2. Strap or tape an electronic thermometer onto each cylinder.
3. Read and review the information provided on the outside of each cylinder.
4. Notice the condition of the cylinder, type of refrigerant and whether the system should be charged with liquid or vapor.
5. Record the following information:

	R134a	401a	404a	R22
pressure	_______	_______	_______	_______
temperature	_______	_______	_______	_______
bubble point	_______	_______	_______	_______
dew point	_______	_______	_______	_______

QUESTIONS
1. Are any of the cylinders marked with arrows to charge upright for liquid? If so, which ones?

2. Does this mean the refrigerant will fractionate if it is charged as a vapor?

3. Which refrigerant(s) do not have bubble point and dew point available on the PT charts?

4. Give a specific application for each refrigerant:

134a __________________________ R22 __________________________

401a __________________________ 404a __________________________

Laboratory Worksheet - # R17
Student____________________________
Instructor ___________________________
Date ____________ Grade__________

COMPRESSOR LUBRICANTS

EXPECTED COMPETENCY:
To demonstrate the ability to understand the applications and considerations of compressor lubricants available for use with various refrigerants

STUDY MATERIAL
Chapter R11

EQUIPMENT
1. Safety goggles
2. Container of polyolester oil
3. Container of alkylbenzene oil
4. Container of mineral oil

SAFETY
Be cautious when working around the containers to avoid contaminating the lubricants.

PROCEDURE
1. Set up the three containers of oils on a work bench and identify the application of each according to the instructions provided by the manufacturer.

Polyolester oil	Alkylbenzene oil	Mineral oil
____________________	____________________	____________________

QUESTIONS
1. Which lubricant is used with HCFC refrigerants?

2. Which lubricant is used with CFC refrigerants?

3. Which lubricant is used with HFC refrigerants?

4. What is the procedure for replacing mineral oil in an existing system with alkylbenzene oil?

5. What is the procedure for replacing mineral oil with polyolester oil in an existing system?

6. Which lubricant is considered highly hygroscopic? (Attracts and retains moisture.)

Laboratory Worksheet - # R18
Student________________________
Instructor _____________________
Date ____________ Grade_________

LEAK DETECTORS

EXPECTED COMPETENCY:
To become acquainted with the various types of non-intrusive leak detectors to detect leaks from systems using modern refrigerants. (Note: a non-intrusive leak detector does not require adding any substance to the refrigerant system, as with the fluorescent dye leak detection methods.)

STUDY MATERIAL
Chapter R6

EQUIPMENT
1. A drum of R-134a, MP39, and R-22
2. A tank of pressurized nitrogen
3. Evaporator coil or condenser coil with known refrigerant leak. 1/4" flare connections at inlet and outlet
4. Two refrigerant hoses with 1/4" female couplings at each end
5. 0-500 pressure gauge with 1/4" male flare connection
6. Leak detectors:
   a. Electronic type, battery or AC power
   b. Liquid bubble solution
   c. Ultrasound
7. Safety goggles

SAFETY
Be cautious when working around refrigerant cylinders to prevent leaking refrigerant. Only *de-minimus* or minimum and unintentional amounts of refrigerant to the atmosphere are legal. Always wear your safety glasses and protective gloves when handling refrigerants.

PROCEDURE
1. Connect the pressure gauge to the outlet of the coil by means of the refrigerant hose.
2. Connect the refrigerant cylinder to the inlet of the coil by means of the refrigerant hose.
3. With your safety goggles on, pressurize the coil to 10 PSIG with the refrigerant and then increase the pressure to 75 pounds with the nitrogen.
4. In turn, use the different types of leak detectors to determine the location of the leak.

Repeat the procedures several times to become thoroughly familiar with the use of leak detectors.

QUESTIONS

1. Which leak detection method was quickest in locating the leak in a general area?

2. Which leak detection method will pinpoint the leak to a specific area?

3. Which leak detector will find a leak using any refrigerant type, CFC, HCFC, or HFC?

4. Will the older CFC electronic leak detectors locate leaks on HFC refrigeration systems?

Laboratory Worksheet - # R19
Student_______________________
Instructor _____________________
Date ____________ Grade_________

REFRIGERANT CONTROL DEVICES-FLOATS

EXPECTED COMPETENCY:
To become acquainted with float-type pressure control devices used in refrigeration systems

STUDY MATERIAL
Chapter R8

EQUIPMENT
1. Samples of cutaways and finger-tight assemblies of:
   a. Low side float
   b. High side float
2. Student set of tools
3. Safety goggles

PROCEDURE
1. Wearing safety goggles, disassemble the high side float.
   a. Note the direction of refrigerant flow through the needle and seat.
   b. Note the direction of refrigerant flow through the float assembly.
2. Reassemble the float, noting the action of the flow.
3. Disassemble the low side float.
   a. Note the direction of refrigerant flow through the needle and seat.
   b. Note the direction of refrigerant flow through the float assembly.

QUESTIONS
1. The action of the high side float is determined by the quantity of refrigerant in the ___________.

2. The action of the low side float is determined by the quantity of refrigerant in the____________.

3. Which type of system is critical of the refrigerant charge?

4. Which type of system uses a receiver?

Laboratory Worksheet - # R20
Student______________________
Instructor ______________________
Date ____________ Grade_________

CAPILLARY TUBE-METERING DEVICES

EXPECTED COMPETENCY:
To become acquainted with the construction and operation of capillary tube metering devices

STUDY MATERIAL
Chapter R8

EQUIPMENT
1. Safety goggles
2. Short roll of capillary tube
3. One capillary tube in operating condition within a refrigeration system
4. Refrigeration gauge manifold
5. Measuring gauge to determine actual inside diameter of capillary tube
6. Source of air pressure

SAFETY
Be cautious when working around the air pressure to prevent over-pressurization and possible explosion of the tubing or hoses.

PROCEDURE
1. Measure the inside diameter of the capillary tube from the short roll and record the size. Inside diameter size _________.
2. Measure the entire length of the capillary tube. Length____________.
3. Solder 1/4" ACR copper tubing to each end of the roll with 1/4" flare nuts on each end.
4. Connect the capillary tube to the gauge manifold with the manifold high pressure gauge and hose to the air pressure supply, the center hose to the cap tube inlet and the compound gauge and hose to the cap tube outlet.
5. Loosen the connection to the low pressure hose so some air can escape when the air is on.
6. Turn on the air to check the connections.
7. Adjust the refrigeration gauges to control the air pressure. What are the inlet and outlet pressures? Inlet_______PSIG Outlet_______PSIG
8. Next, go to the working refrigeration system with an operational capillary tube and record the inlet and outlet pressures and superheat.
   Inlet_______PSIG Outlet_______PSIG Superheat_________°F.

QUESTIONS

1. Can a low pressure control be used to control a refrigeration system using a capillary tube type of metering device?

2. What is the effect of moisture condensing or freezing on the inside of the cap tube? List some remedies.

3. List some outstanding faults of the capillary tube metering device.

4. List some good points of the capillary tube metering device?

5. What information is required when replacing capillary tubes with new ones?

6. What determines the amount of refrigerant that flows through a capillary tube metering device?

Laboratory Worksheet - # R21
Student_________________________
Instructor _______________________
Date ____________ Grade_________

AUTOMATIC EXPANSION VALVE

EXPECTED COMPETENCY:
To become acquainted with the construction and operation of an automatic expansion valve

STUDY MATERIAL
Chapter R8

EQUIPMENT
1. Safety goggles
2. Cutaway samples or finger-tight automatic expansion valve
3. One automatic expansion valve in operating condition
4. Refrigeration gauge manifold
5. Source of air pressure

SAFETY
Be cautious when working around the air pressure to prevent over-pressurization and possible explosion of the tubing or hoses.

PROCEDURE
1. Disassemble the old expansion valve, noting the relationship of the parts to the entire valve. Draw the parts in their relationship on the back of this page. DO NOT copy a valve from the book, but use the valve you have.
2. Re-assemble the valve, making sure that none of the parts are lost and each is in its proper order.
3. Connect the good valve to the gauge manifold with the manifold high pressure gauge and hose to the air pressure supply, the center hose to the valve inlet and the compound gauge and hose to the valve outlet.
4. Loosen the connection to the low pressure hose so that some air can escape when the air is turned on.
5. Turn on the air to check the connections.
6. Adjust the valve by turning it all the way in. What are the inlet and outlet pressures?
   Inlet_______PSIG Outlet_______PSIG
7. Adjust by turning the valve all the way out. What are the inlet and outlet pressures?
   Inlet_______PSIG Outlet_______PSIG
8. Adjust to obtain a constant 10 PSIG. Tighten the bleeder connection. What happens to the outlet pressure?

QUESTIONS

1. Can a low pressure control be used to control a refrigeration system using an automatic expansion valve?

2. What is the effect of moisture condensing on the outside of the valve bellows and freezing there? List some remedies.

3. List some outstanding faults of the automatic expansion valve.

4. List some good points of the automatic expansion valve.

5. What is the effect on the operation of the refrigeration system of too high a valve setting?

6. How does the automatic expansion valve respond to a load increase?

7. The automatic expansion valve maintains a constant ______________.

8. The automatic expansion valve feeds (more or less) refrigerant with an increase in load.

Laboratory Worksheet - # R22
Student____________________________
Instructor ___________________________
Date _____________ Grade__________

THERMOSTATIC EXPANSION VALVES

EXPECTED COMPETENCY:
To become acquainted with the construction and operation of a thermostatic expansion valve

STUDY MATERIAL
Chapter R8

EQUIPMENT
1. Safety goggles
2. Cutaway samples or finger-tight thermostatic expansion valve
3. One thermostatic expansion valve in operating condition
4. Refrigeration gauge manifold
5. Source of air pressure

SAFETY
Be cautious when working around the air pressure to prevent over-pressurization and possible explosion of the tubing or hoses.

PROCEDURE
1. Disassemble the old expansion valve, noting the relationship of the parts to the entire valve. Draw the parts in their relationship on the back of this page. DO NOT copy a valve from the book, but use the valve you have.
2. Re-assemble the valve, making sure that none of the parts are lost and that each is in its proper order.
3. Connect the valve to the gauge manifold with the manifold high pressure gauge and hose to the air pressure supply, the center hose to the valve inlet and the compound gauge and hose to the valve outlet.
4. Loosen the connection to the low pressure hose so that some air can escape when the air is on.
5. Insert the sensing bulb in a jar of ice water.
6. Turn on the air to check the connections.
7. Adjust the valve superheat by turning the valve stem halfway in. What are the inlet, outlet pressures, and superheat? Inlet_______PSIG Outlet_______PSIG Superheat_________
8. Adjust the valve superheat by turning the valve stem halfway out. What are the inlet, outlet pressures, and superheat? Inlet_______PSIG Outlet_______PSIG Superheat_________

9. Adjust to constant 10°F. superheat, tighten the bleeder connection. What happens to the outlet pressure?
10. Repeat the above process by removing the sensing bulb from the ice water and let it sense ambient temperature.
11. Adjust the valve superheat by turning the valve stem halfway in. What are the inlet, outlet pressures, and superheat? Inlet_______PSIG Outlet_______PSIG Superheat_________
12. Adjust the valve superheat by turning the valve stem halfway out. What are the inlet, outlet pressures, and superheat? Inlet_______PSIG Outlet_______PSIG Superheat_________
13. Adjust to constant 10°F. superheat, tighten the bleeder connection. What happens to the outlet pressure?

QUESTIONS

1. Can a low pressure control be used to control a refrigeration system using a thermostatic expansion valve?

2. What is the effect of moisture condensing on the outside of the valve bellows and freezing there? List some remedies.

3. List some faults of the thermostatic expansion valve.

4. List some good points of the thermostatic expansion valve.

5. What is the effect on the operation of the refrigeration system of too high a superheat setting?

6. What is the effect on the operation of the refrigeration system of too low a superheat setting?

7. When a coil is starved for refrigerant, the superheat is _____________.

8. When a coil is flooded with refrigerant, the superheat is _____________.

9. A thermostatic expansion valve maintains a constant __________________.

10. The thermostatic expansion valve feeds (more or less) refrigerant with an increase in load.

11. What are the three forces that affect the TXV ?

12. Where is the bulb of the TXV mounted?

13. Why do some TXV's require an external equalizer line?

14. Where is the external equalizer line connected into a system?

Laboratory Worksheet - # R23

Student___________________________

Instructor ________________________

Date ____________ Grade__________

ELECTRONIC-SOLID STATE EXPANSION VALVES

EXPECTED COMPETENCY:

To become acquainted with the construction and operation of electronic expansion valves

STUDY MATERIAL

Chapter R8

EQUIPMENT

1. Safety goggles
2. Cutaway samples or finger-tight electronic expansion valve
3. One electronic expansion valve in operating condition
4. One electronic expansion valve installed in a working refrigeration system
5. Electronic control module or computer interface with superheat temperature sensor located at outlet of evaporator
6. Refrigeration gauge manifold
7. Source of air pressure
8. Student tool box

SAFETY

Be cautious when working around the electronic valve to avoid creating an electrical short and possibly destroying the valve. Also, be careful to prevent over-pressurization and possible explosion of the tubing or hoses.

PROCEDURE

1. Disassemble the old expansion valve, noting the relationship of the parts to the entire valve. Draw the parts in their relationship on the back of this page. DO NOT copy a valve from the book, but use the valve you have.
2. Re-assemble the valve, making sure that none of the parts are lost and that each is in its proper order.
3. Connect the good valve to the gauge manifold with the manifold high pressure gauge and hose to the air pressure supply, the center hose to the valve inlet and the compound gauge and hose to the valve outlet.
4. Loosen the connection to the low pressure hose so that some air can escape when the air is on.
5. Turn on the air to check the connections.
6. Adjust the valve superheat via the controller or computer to modulate the valve half a revolution in. What are the inlet and outlet pressures?

   Inlet_______PSIG Outlet_______PSIG

7. Adjust the valve superheat via the controller or computer to modulate the valve half a revolution out. What are the inlet and outlet pressures? Inlet________PSIG Outlet________PSIG
8. Adjust to a constant 10 _F. superheat, tighten the bleeder connection. What happens to the outlet pressure?
9. Repeat the above process with the electronic expansion valve in the working refrigeration system. Be sure to install the refrigeration gauges on the high and low sides of the system. Note: the high side of the refrigeration system is found at the compressor discharge line (the smaller tube) and the suction side is found at the compressor suction line (the larger tube). If you are unfamiliar with using compressor service valves, ask your instructor for assistance to demonstrate the proper operation.
10. Adjust the valve superheat to 5 degrees superheat. What are the inlet and outlet pressures?
Inlet________PSIG Outlet________PSIG
11. Adjust the valve superheat on the controller board to 10 degrees superheat. What are the inlet and outlet pressures?
Inlet________PSIG Outlet________PSIG
12. With the working refrigeration system, install the superheat sensing element in ice water and record the inlet and outlet pressures.
Inlet________PSIG Outlet________PSIG
13. Now let the superheat sensing element sense the ambient air temperature. Record the inlet and outlet pressures.
Inlet________PSIG Outlet________PSIG

QUESTIONS

1. Can a low pressure control be used to control a refrigeration system using an electronic expansion valve?

2. What is the effect of moisture condensing on the valve and freezing there?

3. List some outstanding faults of the electronic expansion valve.

4. List some good points of the electronic expansion valve.

5. What type of sensor is used to measure the superheat of the evaporator?

6. The control signal that goes to the valve to modulate open and close is what type of electronic signal?

7. What process would be used to troubleshoot this type of valve?

Laboratory Worksheet - # R24

Student____________________________

Instructor __________________________

Date _____________ Grade__________

DISASSEMBLE AND RE-ASSEMBLE AN OPEN TYPE COMPRESSOR

EXPECTED COMPETENCY:

To become acquainted with the construction and operation of a compressor

STUDY MATERIAL

Chapter R11

EQUIPMENT

1. Open type compressor--belt driven
2. Safety goggles
3. Student tool kit
4. Work bench with vise
5. Torque wrench with assorted sockets
6. Flywheel puller
7. Gauge manifold set with hoses
8. Compressor test bench
9. Oil pan
10. Parts pan
11. Electronic leak detector

SAFETY

Be cautious when removing the flywheel and older components that are stuck and hard to remove. When working around the nitrogen pressure, be sure not to over-pressurize the system. Be sure to use a nitrogen source with a properly working regulator and don't allow your operating pressure to exceed 50 psi.

PROCEDURE

1. Remove the flywheel by:
   a. Loosening the flywheel nut two turns - DO NOT REMOVE IT.
   b. Use flywheel puller to loosen flywheel on crankshaft. Retainer nut will prevent flywheel from leaving shaft and being damaged.
   c. Remove nut, flywheel and shaft key. Put parts in pan. Remember how they came off.
2. Remove oil port plug and pour any oil in the crankcase into a metal container. PUT PLUG IN PARTS PAN.

3. With compressor clamped firmly but not excessively in the bench vise, loosen all the bolts holding the head and valve plate. After all the bolts have been loosened, remove and put into parts pan.
4. Remove head and valve plate. If the gasket holds the head or valve plate on, DO NOT HIT THE PART WITH A HAMMER. Using a small cold chisel at the edge of the gasket, lightly tap the chisel to loosen the part. When assembling the compressor, new gaskets must be used.
5. Inspect the head to see the circuit of refrigerant flow from suction to discharge.
6. Remove any nut or bolt head retainers on the valve plate. Remove the holding bolts or retainers holding the discharge valves. Notice the positions of retainers and valve loading springs.
7. Re-assemble the valve plate.
8. Put valve plate and suction valves in parts pan.
9. Invert the compressor body and loosen the crankcase cover screws.
10. Remove the screws and crankcase cover. The gasket will usually stick and the chisel and a light tap of the hammer are required.
11. Turn the crankshaft to note the action of the shaft, connecting rods and piston.
12. Before removing the connecting rod caps, mark them to identify their location and direction. The cap and rod must be re-assembled in exactly the same relationship. This is done with a prick punch on the connecting rod cap on the flywheel side of the cap. One punch mark for the cap closest to the flywheel, two for the next cap, etc.
13. Turn the assembly with pistons up and mark the pistons in the same manner. Only a light mark is needed.
14. Loosen and remove the connection rod, bolt retainers and bolts. Bolt retainers are of many different styles. Check with the instructor on the proper way of removing the retainers from the compressor you are working on.
15. Remove the connecting rod caps and inspect for even wear. Consult with the instructor on the wear encountered.
16. Push the piston-rod assembly out the cylinder top.
17. Check the wear of the piston pin by wiggling the rod in the piston. No side movement should be felt.
18. Remove the seal plate from the flywheel end of the shaft and pull the seal assembly. Be careful not to break the carbon-seal face.
19. Pull the crankshaft from the housing.
20. Check and identify each part and review its position as compared to the other parts.
21. Note the method of lubrication used. It will be either an oil pump on the end of the shaft, or the splash method which scoops oil up onto the connecting rod cap.
22. Re-assemble the compressor assembly in the reverse order of the disassembly process.
    a. Insert crankshaft.
    b. Install seal assembly and seal plate with a new gasket. A trick to remember: the use of a refrigerant oil coating on the gasket will help ensure a good seal.
    c. Insert the seal plate bolts until finger-tight.
    d. Tighten with a torque wrench in a cross pattern only, 1/4 turn at a time. Cross pattern sequence is using the clock face, 12, 6, 3, 9, 1, 7, 4, 10, 2, 5, 8, 11, 12, etc. Repeat the clock pattern until all bolts are torqued to the proper compressor specifications.
    e. Check for crankshaft binding by turning it.

f. Place piston connection rod assemblies in their proper cylinders with the identification marks toward the flywheel end of the crankshaft.
g. Place connecting rod caps on the proper connecting rods according to the marks on the caps. Insert the bolts and retainers.
h. Tighten the bolts in a cross pattern, 1/4 turn on each bolt until proper torque is obtained.
i. Turn the shaft by hand to make certain no caps or piston pins are binding.
j. Re-install crankcase cover. Tighten bolts using cross pattern.
k. Re-install the valve plate, suction valves and head in the correct position to obtain proper refrigerant flow from the suction intake to the discharge.

23. Put in the proper amount of refrigerant oil according to the specifications for the compressor. Usually this is to a level 1" below the edge of the oil plug opening. Tighten the oil plug to proper torque specifications.
24. Re-install the flywheel with key and nut. Tighten the nut to the proper specifications.
25. Rotate the flywheel several times to put oil into the upper parts of the compressor body.
26. Leak test the assembly.
    a. Connect the gauge manifold to the suction and discharge gauge ports. Front seat the service valve to seal off the compressor. This is done by turning the valve clockwise until it stops.
    b. Using a drum of R-22, pressurize the compressor through the center hose to 10 PSIG and then to 50 PSIG with nitrogen pressure. Check all assembled parts for leaks around gaskets as well as bolt heads, using your electronic leak detector.
    c. Relieve the pressure through the center hose to atmosphere. (Note: this is a *de minimus* amount.)
27. Install the compressor assembly on the test bench and test for vacuum and pressure. This procedure will vary with the make and style of compressor - check with the instructor.

QUESTIONS

1. Compressors having two or more cylinders have the connecting rods connected to the crankshaft at different angles. Why?

2. Why should the clockwork pattern of tightening bolts be used?

3. A medium or low temperature refrigeration compressor should be able to pull to what vacuum?

4. An air conditioning compressor will not be able to pull as deep a vacuum as a medium or low temperature compressor. Why?

5. How long should the compressor be able to hold the crankcase vacuum?

Laboratory Worksheet - # R25

Student__________________________

Instructor ________________________

Date ____________ Grade_________

SEMI-HERMETIC COMPRESSOR ASSEMBLY

EXPECTED COMPETENCY:
To become acquainted with the construction and operation of a serviceable compressor assembly

STUDY MATERIAL
Chapter R11

EQUIPMENT
1. Semi-hermetic type motor-compressor assembly
2. Safety goggles
3. Student tool kit, including torque wrench and assorted sockets
4. Gauge manifold and hoses
5. Drum R-22
6. Leak detector
7. Parts pan
8. Refrigerant oil (compatible with system)

PROCEDURE
1. Remove oil plug and pour oil from compressor crankcase.
2. Using the techniques learned in laboratory task R24, remove the head and valve plate assembly, crankcase plate, oil pump plate and motor end bell.
3. Rotate the crankshaft and note the action of the piston assemblies, as well as the lubrication methods. The method used to provide lubrication could be splash, positive pressure with oil pump or dip arms and catch cup.
4. Replace the plates using new oiled gaskets and proper bolt-tightening techniques.
5. Using gauge manifold and hoses and refrigerant supply, pressurize the compressor assembly and check for leaks.
6. Recover/recycle the refrigerant from the compressor assembly. Note: Refer to Lab Worksheet #R33 for refrigerant recovery procedures.
7. Connect the compressor to the proper power supply. Check with instructor.
8. Operate the compressor for vacuum test.

QUESTIONS
1. Is there any difference in pumping action between the semi-hermetic compressor and the open type?

2. What is the difference between the semi-hermetic compressor and the open type?

3. How does the oil returning from the system reach the motor end of the assembly?

4. Small semi-hermetic compressors are built to operate at what two speeds?

Laboratory Worksheet - # R26
Student___________________________
Instructor ________________________
Date _____________ Grade__________

HERMETIC TYPE COMPRESSOR

EXPECTED COMPETENCY:
To become acquainted with the construction and operation of a hermetic type motor-compressor assembly

STUDY MATERIAL
Chapter R11

EQUIPMENT
1. Safety goggles
2. Cutaway model - hermetic motor-compressor assembly

PROCEDURE
1. Trace the flow of the refrigerant from the inlet connection on the dome to the discharge outlet connection.
2. Identify each part and explain its purpose for the operation of the compressor.

QUESTIONS
1. What is the purpose for the assembly at the top of the motor stator?
2. What is the purpose of the assembly at the outlet of the compressor discharge?
3. Why does the hot gas line travel around the inside of the shell before being connected to the outlet connection?
4. What type of insulation is used on the motor leads?
5. How are the motor terminals sealed into the shell?
6. Does the motor have overload protection? If so, what type is it?
7. How is this type of compressor motor normally cooled?

8. How is the typical hermetic compressor mounted inside the shell?

9. Is the internal compressor shell on the high or low side pressure of the refrigeration system?

10. Which compressor valves are larger, the suction or discharge?

Laboratory Worksheet - # R27
Student____________________________
Instructor ____________________________
Date ______________ Grade___________

ROTARY COMPRESSOR

EXPECTED COMPETENCY:
To become acquainted with rotary type compressors

STUDY MATERIAL
Chapter R11

EQUIPMENT
1. Safety goggles
2. Rotary compressor cutaway
3. Rotor assembly
4. Student tool kit

PROCEDURE
1. Using the rotary compressor cutaway, trace the refrigerant circuit from the suction intake to the cylinder, through the chambers of the cylinder to the discharge in the shell and to the outlet of the assembly.
2. Remove the end bearing plate of the rotor assembly. Trace the refrigerant vapor flow through the chamber by the action of the eccentric and blade.

QUESTIONS
1. How does the action of a rotary compressor differ from a reciprocating compressor?

2. Are valves used in the rotary compressor to produce the compressing action?

3. What is the purpose of the check valve in the cylinder outlet?

4. Does the rotary compressor need an oil pump? Why or why not?

5. What is the purpose of the large cylinder mounted on the side of the compressor in the discharge circuit?

Laboratory Worksheet - # R28

Student____________________

Instructor ____________________

Date ____________ Grade__________

SCROLL COMPRESSORS

EXPECTED COMPETENCY:

To become acquainted with scroll type compressors

STUDY MATERIAL

Chapter R11

EQUIPMENT

1. Safety goggles
2. Scroll compressor cutaway
3. Student tool kit

PROCEDURE

1. Using the scroll compressor cutaway, trace the refrigerant circuit from the suction intake to the cylinder, through the scroll inside the cylinder to the discharge in the shell and to the outlet of the assembly.
2. Remove the end bearing plate of the compressor assembly. Trace the refrigerant vapor flow through the chamber by the action of the eccentric scroll and housing.

QUESTIONS

1. How does the action of a scroll compressor differ from a reciprocating compressor?

2. Are there any types of bearings used in the scroll compressor? If so, what are they?

3. What is the purpose of the check valve in the cylinder outlet?

4. Does the scroll compressor need an oil pump? Why?

5. Why is this type of compressor more efficient than conventional reciprocating compressors?

6. The scroll compressor is very quiet during normal operation. Why?

7. What process would be used to troubleshoot the scroll compressor?

8. Why is a short cycle relay used with some applications of scroll compressors?

Laboratory Worksheet - # R29
Student________________________
Instructor ________________________
Date ____________ Grade__________

CONDENSERS - SHELL AND COIL

EXPECTED COMPETENCY:
To become acquainted with the proper method of cleaning accessible shell and tube condensers

STUDY MATERIAL
Chapter R10

EQUIPMENT
1. Safety goggles
2. Shell and tube condensers
3. Student tool kit
4. Tube cleaning rod
5. Electric drill

SAFETY
Be cautious when working around water and electricity. Be sure the electric drill is properly grounded to avoid electrical shock.

PROCEDURE
1. Mark the side of the condenser heads and shell body to ensure exact replacement for proper water circuiting. DO THIS BEFORE REMOVING THE HEADS.
2. Remove the heads and gaskets. Note the positions of the gaskets. They or new ones must go on in the same position.
3. Inspect the water tubes for scale deposits.
4. Using the tube cleaning rods in the electric drill, ream out the tubes. Not all the deposits can be removed with the reamers. To finish the cleaning, wire brush rods the length of the tubes are used.
5. Re-install the heads and gaskets. BE SURE THEY GO ON IN THE SAME POSITION AS BEFORE THEY WERE REMOVED.
6. Connect the water circuit of the condenser to a water source and check for leaks.

QUESTIONS
1. Where is the greatest buildup of deposits in the condenser, at the refrigerant inlet or outlet? Why?

2. Why does the circuiting of the condenser produce counterflow of water to refrigerant?

3. What would be the effect of having the gasket out of position?

4. Why do some condensers have to be cleaned with brushes and others with chemicals?

5. When is the most heat removed from the refrigerant in the condensing process?

Laboratory Worksheet - # R30

Student___________________________

Instructor ________________________

Date ____________ Grade__________

CONDENSER - CO-AXIAL

EXPECTED COMPETENCY:

To become acquainted with the proper method of cleaning a tube-in-a-tube or coaxial type condenser

STUDY MATERIAL

Chapter R10

EQUIPMENT

1. Safety goggles
2. Rubber gloves and rubber apron
3. Water-cooled unit using tube-in-a-tube condenser
4. Student tool kit
5. Acid pumping kit

SAFETY

Be cautious when working around the acid cleaning solution so that you do not splash or spill it. (The consequences of a spill will result in expensive cleanup procedures as per OSHA or local regulations.)

PROCEDURE

1. Safety goggles, rubber gloves and rubber apron must be worn at all times when using acid.
2. Connect the acid pumping kit to the inlet and outlet of the condenser. A properly installed unit will have connecting valves and isolating valves for this purpose.
3. Mix the cleaning solution by:
   a. Filling the acid container half full of water.
   b. Pour the required amount of cleaning acid into the water. The solution must be formed by pouring the acid into the water. NEVER pour water into the acid - an explosion could result.
4. Run the pump forcing acid solution through the condenser and back into the container.
5. Continue this operation until the foaming ceases in the solution returning from the condenser.
6. Dispose of the acid solution and fill the container three-fourths full of water.
7. Run the pump until the solution comes out clear.
8. Remove the acid kit hoses and put the unit back into operation.

QUESTIONS

1. Does it make any difference which direction the acid solution flows through the condenser?

2. How can you tell when the condenser tubes are clean?

3. Why is it necessary to flush the acid solution out of the condenser tubes before putting the condenser back on the operating water supply?

4. What are the environmental requirements for your state when using acid solution to clean the condenser tubes?

Laboratory Worksheet - # R31
Student__________________________
Instructor __________________________
Date _____________ Grade__________

REFRIGERANT PIPING

EXPECTED COMPETENCY:
To become acquainted with the sizing and installation of refrigerant piping

STUDY MATERIAL
Chapter R16

EQUIPMENT
1. Safety goggles
2. Evaporator blower unit or air handler
3. Condensing unit
4. Quantity of copper tubing of determined size
5. Student tool kit
6. Pipe insulation of determined size
7. Low temperature silver solder
8. Flux
9. Oxygen/acetylene welding equipment
10. Flare nuts and fittings
11. Refrigerant supply drum
12. Leak detectors
13. Refrigerant gauge set
14. Nitrogen tank

PROCEDURE
1. Determine the size of pipes involved according to procedures in Chapter 13.
2. Fabricate the suction line as required according to the equipment assigned.
3. Install pipe insulation on sections of suction line.
4. Solder lines together with appropriate fittings.
5. Fasten suction line to DX coil and condensing unit with appropriate fittings.
6. Install liquid line with proper bends and fittings to parallel the suction line.
7. Connect the liquid line to the DX coil pressure reducing device with appropriate fittings.
8. Using the refrigerant gauge set connected to the service connections on the condensing unit, pressurize the refrigerant lines to 10 PSIG and then to 50-60 PSIG using nitrogen.
9. Check for leaks using the leak detectors supplied from tool room.
10. Have instructor check your results.

QUESTIONS

1. Horizontal lines should pitch in what direction? How much?

2. Why should the suction line be insulated?

3. Is there any situation where the liquid line should be insulated?

4. When sizing the suction line, the minimum velocity of vapor flow is ___ FPM for horizontal lines and ___FPM for vertical lines.

5. The maximum pressure drop through the suction line is ___PSIG for R-401a , ___PSIG for R-22 and ___PSIG for R-404.

6. What is the effect on the refrigeration system if too much pressure drop exists on the suction line?

7. Where are traps required in the suction line?

8. Is insulation ever required on the hot gas line?

9. What is the maximum pressure drop allowable in the liquid line?

10. What would be the effect on the refrigeration system of too great a pressure drop in the liquid line?

11. What is the effect of elevation on the liquid line when the DX coil is above the condensing unit?

12. What is the effect of elevation on the liquid line when the condensing unit is above the DX coil?

Laboratory Worksheet - # R32
Student___________________________
Instructor ________________________
Date _____________ Grade__________

REFRIGERATION PRESSURE MEASUREMENTS

EXPECTED COMPETENCY:
To become acquainted with the use of the gauges and manifold set

STUDY MATERIAL
Chapters R17 and R18

EQUIPMENT
1. Safety goggles
2. Gauge manifold set with hoses (three-way valve type)
3. Drum of R-22 refrigerant
4. Student tool set

SAFETY
Be cautious when working around the refrigerant not to release any to the atmosphere, except a *de minimus* or minimum amount.

PROCEDURE
Wearing safety goggles, proceed as follows:
1. With manifold valves closed and flare plug in the center hose connection, connect the refrigerant cylinder to the compound gauge hose. When the cylinder valve is opened, note any change in the compound and pressure gauge readings.
2. With the cylinder connected to the pressure gauge hose, open both valves on the manifold.
3. When the cylinder valve is opened, note the gas flow that takes place.
4. With the refrigerant cylinder connected to the center hose and the cylinder valve opened, open the manifold valves and note the operation.

QUESTIONS
1. The valve beneath each gauge controls the pressure to the gauge. True or False?

2. With the manifold connected to an operating unit, if both valves are opened, how would the operation of the unit be affected?

3. In what service procedure would both manifold valves be opened?

4. When purging *de minimus* refrigerant from the gauge manifold, which valve(s) would be opened?

5. When adding refrigerant vapor to the low side of the system, which manifold valve would be opened?

6. What would be the possible result of adding liquid refrigerant into the low side of the system?

Laboratory Worksheet - # R33
Student_______________________
Instructor _____________________
Date ___________ Grade_________

REFRIGERANT RECOVERY/RECYCLE

EXPECTED COMPETENCY:

To demonstrate the ability to safely recover and recycle refrigerant from two refrigeration systems using two different types of recovery equipment

STUDY MATERIAL

Chapter R14

EQUIPMENT

1. Safety goggles
2. Two different refrigerant recovery/recycle machines (if available)
3. Two working refrigeration systems
4. Student tool box
5. Acid test kit
6. Oil storage container

SAFETY

Be cautious when working around the refrigerants. There may be residual acids contained in the oil. Be sure to wear safety goggles and hand protection when working with any refrigerants. Don't store refrigerant recovery cylinder in direct sunlight; this may over-pressure the cylinders.

PROCEDURE

1. Install gauges on the refrigeration equipment.
2. Connect the hoses from the refrigerant recovery machine to the refrigeration system. *Note: Be sure to read and follow the recovery machine manufacturer's instructions prior to the installation.
3. Weigh the refrigerant recovery tank.
4. Begin recovering the refrigerant.
5. When this process is complete, record the following:

	System #1	System #2
Weight of recovery cylinder before recovery	___________	___________
Weight of recovery cylinder after recovery	___________	___________
Amount of time to recover the refrigerant	___________	___________
Amount of oil recovered from the system	___________	___________

6. Begin recycling the refrigerant.
7. When this process is complete, test the recycled refrigerant for acids, using a portable acid test kit.

8. Repeat the above process for refrigeration system #2 with a different type of recovery machine.
9. Put away all tools, and replace covers on equipment.

QUESTIONS

1. Which recovery/recycle machine was able to remove the refrigerant at a faster rate? Why?

2. Did the oil acid test indicate potential acids? If so, what is the procedure to clean up the acids from the refrigerant?

3. How deep of a vacuum did the recovery machines pull on the system before they shut down?

4. Did you recover as a liquid or a vapor, or both? Explain each machine and the process.

5. Is it possible to also reclaim the refrigerant with the same recovery/recycle machine when working in the field?

6. After a hermetic motor burnout, what contaminant might be found in the recovered refrigerant?

7. What would the result be if a recovery cylinder were filled with liquid and allowed to warm up?

Laboratory Worksheet - # R34
Student______________________
Instructor ____________________
Date ___________ Grade________

SYSTEM EVACUATION

EXPECTED COMPETENCY:
To become acquainted with the methods of evacuating refrigeration systems

STUDY MATERIAL
Chapter R18

EQUIPMENT
1. Safety goggles
2. Refrigeration system that has service valves for gauge connections
3. Gauge manifold set
4. Drum of refrigerant that is used in the unit
5. Student tool kit
6. High vacuum gauge

SAFETY
When using the vacuum pump, be careful not to pressurize the pump with refrigerant pressure. This may cause permanent damage and void any vacuum pump manufacturer's warranties.

PROCEDURE
1. Follow the procedure outlined in text book to evacuate the system by the deep vacuum method. Note the time involved in each step. Record the total time involved.
2. Follow the procedure outlined in text book to evacuate the system by the triple evacuation method. Note the time involved in each step. Record the total time involved.

QUESTIONS
1. Which method of evacuation required the least time on the small horsepower unit?
2. Which method of evacuation would be the most practical on a 200 H.P. unit?
3. What is the purpose of system evacuation?
4. What happens to a system that leaks while in a vacuum?

5. Why is vacuum oil used in a vacuum pump instead of refrigeration oil?

6. Define a deep vacuum.

7. What vapors does a deep vacuum pull out of a system?

8. Name two common vacuum test gauges.

Laboratory Worksheet - # R35

Student__________________________

Instructor ________________________

Date ____________ Grade_________

REFRIGERANT CHARGING

EXPECTED COMPETENCY:

To become acquainted with the methods of charging the refrigeration system

STUDY MATERIAL

Chapter R18

EQUIPMENT

1. Safety goggles
2. Small refrigeration system
3. Gauge manifold set
4. Cylinder of refrigerant used in the system
5. Refrigerant scale
6. Student tool kit
7. Pail or tub of hot water

SAFETY

Wear your safety goggles at all times when working around refrigerants and be careful not to vent any refrigerant into the atmosphere.

PROCEDURE

A. Vapor method of refrigerant charging:

****NOTE OF CAUTION****When using some refrigerants with a high glide, the refrigerant will fractionate during charging if charged as a vapor, e.g., MP39 (401a). These new refrigerants have opposite charging procedures when compared with R-22. The cylinder is right side up for liquid charging and upside down for vapor charging. Only vapor charge with these refrigerants if the entire cylinder is to be used. Consult with your instructor for further details.

1. To ensure a sufficient amount of heat in the cylinder of refrigerant to remove the required amount of refrigerant, the cylinder must be heated BY SETTING IN WARM WATER UNTIL THE CYLINDER PRESSURE REACHES THE EQUIVALENT OF 125°F. DO NOT HEAT WITH TORCH OR SET ON OPEN FLAME - CYLINDER COULD EXPLODE.
2. From the data plate of the unit, determine the type and quantity of refrigerant charge.
3. Connect the gauges to the system and evacuate the system following the procedure outlined in the text book.
4. After charging operation is completed, check the unit for proper operation.

5. Charge the system using the procedure outlined in text book.

B. Liquid method of refrigerant charging:

1. Liquid charging does not require vaporizing the refrigerant in the cylinder; therefore, normal cylinder temperatures of 70°F to 80°F are usually sufficient to complete the charging operation.
2. From the data plate on the unit, determine the type and amount of refrigerant charge.
3. To accurately charge a system, the system must be empty. (No refrigerant vapor - at a deep vacuum). Recover any refrigerant pressure in the system and evacuate according to the procedure outlined in text book.
4. Liquid charge the system according to the procedure outlined in text book.
5. After the system is charged, check for proper operation.

QUESTIONS

1. When vapor charging, does the refrigerant go into the discharge or suction side of the compressor?

2. Is the compressor operating or off when vapor charging?

3. What is the major advantage of the vapor charging method?

4. What is the major disadvantage of the vapor charging method?

5. When liquid charging, does the refrigerant go into the discharge or suction side of the compressor?

6. What are the major advantages and disadvantages of liquid charging?

7. Is it possible to accurately, partially charge a system by the liquid charging method?

8. Why is the line to the refrigerant cylinder purged before charging?

9. Why would the cylinder pressure drop when charging vapor into a system?

10. What would happen if liquid refrigerant were to enter the compressor while it was running?

Laboratory Worksheet - # R36
Student____________________________
Instructor __________________________
Date _____________ Grade_________

DOMESTIC REFRIGERATORS

EXPECTED COMPETENCY:
To demonstrate the ability to maintain, service, and troubleshoot domestic refrigerators

STUDY MATERIAL
Chapter DR22

EQUIPMENT
1. Safety goggles
2. 1 frost free domestic refrigerator
3. 1 manual defrost domestic refrigerator
4. Refrigeration gauges
5. Hermetic compressor analyzer
6. 3 electronic thermometers
7. VOM with amp clamp
8. Student tool box

SAFETY
Be cautious when working with pressures on the high side of a domestic refrigerator. They can reach pressures in excess of 125 PSIG. Use quick connect refrigeration fittings to minimize the loss of refrigerant.

PROCEDURE
1. Install refrigeration line tap fittings (if required).
2. Install refrigeration gauges.
3. Connect amp clamp meter to common wire on the compressor.
4. Insert temperature probes into refrigerator and freezer sections.
5. Measure and record the following:

	Frost free unit	Manual defrost unit
Suction pressure	________________	________________
Discharge pressure	________________	________________
Compressor amps	________________	________________
Superheat	________________	________________
Refrigerator box temperature	________________	________________
Freezer box temperature	________________	________________

6. Become familiar with the electrical schematic provided with each unit.
7. Ask your instructor to install a problem symptom into each unit.

8. Test, repair and record your conclusions.
9. Upon completion remove all test instruments and equipment.

QUESTIONS

1. What is the normal suction pressure for a refrigerator when the freezer is operating at 0°F?
2. What is the normal discharge pressure for a refrigerator when the ambient temperature is 75°F?
3. What were the symptoms and actual problem with the frost free unit?
4. What were the symptoms and actual problem with the conventional unit?
5. How could the compressor analyzer be used to test the compressor?
6. Why should gauges be installed only when absolutely necessary on a domestic refrigerator?
7. What would the symptoms be for a current relay if the contacts were stuck shut?
8. Do all refrigerators have start relays?
9. What is the purpose of the green wire in a three-wire power cord?
10. How would high head pressure affect the operation of the refrigerator?

Laboratory Worksheet - # R37
Student________________________
Instructor ______________________
Date ____________ Grade__________

ICE MACHINES - CUBERS

EXPECTED COMPETENCY:
To demonstrate the ability to startup, test, and troubleshoot cuber ice machines

STUDY MATERIAL
Chapter CR25

EQUIPMENT
1. Safety goggles
2. A cuber ice machine
3. Refrigeration gauges with quick connect/disconnect fittings
4. VOM meter with amp clamp
5. Electronic thermometers

SAFETY
Be cautious when working around the high side gauge fittings on ice machines. If they are a Schrader-type fitting, use quick connect/disconnect gauge fittings to avoid losing any refrigerant. Since these machines are considered a critical charge, any loss of refrigerant will affect operation.

PROCEDURE
1. Install refrigerant gauges on suction and discharge lines.
2. Install thermometers on the suction line and liquid.
3. Locate the electrical junction box and identify the wiring going to the compressor. Install clamp amp meter to the common wire to the compressor.
4. Turn on the ice machine and record the following:

	At startup	During harvest	Immediately after harvest
Compressor amps	________	________	________
Suction pressure	________	________	________
Discharge pressure	________	________	________
Superheat	________	________	________
Subcooling	________	________	________

5. Record the time of each complete cycle from start to finish.
6. Locate the service manual for the ice machine. Become familiar with troubleshooting procedures and test requirements.
7. Ask your instructor to install a problem symptom into the unit.
8. Test, repair and record your conclusions.

9. Upon completion remove all test instruments and equipment.
10. Re-install all covers on the unit.

QUESTIONS

1. Why did compressor amps change right before defrost and after defrost?

2. What control method is used to shut down the ice machine when the bin is full?

3. Did superheat change during the harvest cycle? If so, why?

4. Did the subcooling temperature change during the harvest cycle? If so, why?

5. List the safety controls and the normal operating controls found on the unit.

6. What method was used to initiate and terminate the defrost cycle?

7. Explain the problem you encountered when you were troubleshooting the ice machine. Also explain the solution or remedy to fix the problem.

8. How is cubed ice formed and made?

9. What is the appropriate evaporator temperature when ice harvests in a cube ice maker?

10. How is ice kept refrigerated in the holding bin of a typical ice maker?

Laboratory Worksheet - # R38
Student______________________
Instructor ______________________
Date ____________ Grade________

ICE MACHINES - FLAKERS

EXPECTED COMPETENCY:
To demonstrate the ability to startup, test, and troubleshoot flaker ice machines

STUDY MATERIAL
Chapter CR25

EQUIPMENT
1. Safety goggles
2. A flaker ice machine
3. Refrigeration gauges with quick connect/disconnect fittings
4. VOM meter with amp clamp
5. Electronic thermometers

SAFETY
Be cautious when working around the high side gauge fittings on ice machines. If they are a Schrader-type fitting, use quick connect/disconnect gauge fittings to avoid losing any refrigerant. Since these machines are considered a critical charge, any loss of refrigerant will affect operation.

PROCEDURE
1. Install refrigerant gauges on suction and discharge fittings.
2. Install thermometers on the suction line and liquid line.
3. Locate the electrical junction box and identify the wiring going to the compressor. Install clamp amp meter to the common wire to the compressor.
4. Turn on the ice machine and record the following:

	At startup	During harvest
Compressor amps	__________	__________
Suction pressure	__________	__________
Discharge pressure	__________	__________
Superheat	__________	__________
Subcooling	__________	__________

5. Record the time to begin producing flaked ice.
6. Locate the service manual for the ice machine. Become familiar with troubleshooting procedures and test requirements.
7. Ask your instructor to install a problem symptom into the unit.
8. Test, repair and record your conclusions.

9. Upon completion remove all test instruments and equipment.
10. Re-install all covers on the unit.

QUESTIONS

1. What are the primary differences between the cuber ice machine and the flaker ice machine?

2. What control method is used to shut down the ice machine when the bin is full?

3. Does superheat change during the harvest refrigeration cycle?

4. Does the subcooling temperature change during the harvest refrigeration cycle?

5. List the safety controls and the normal operating controls found on the unit.

6. Explain the problem you encountered when you were troubleshooting the ice machine. Also explain the solution or remedy to fix the problem.

7. What would be the symptoms of a defective float on a flaker ice machine?

8. How is flaked ice made?

Laboratory Worksheet - # R39
Student___________________________
Instructor ________________________
Date ___________ Grade__________

TROUBLE-SHOOT A REFRIGERATION SYSTEM

EXPECTED COMPETENCY:
To become acquainted with the steps to obtain the necessary information to diagnose trouble

STUDY MATERIAL
Chapter TR48

EQUIPMENT
1. Safety goggles
2. Student tool kit
3. Gauge manifold set
4. Six thermometers (pocket type or digital)
5. Clamp type volt-ammeter
6. Refrigeration or air conditioning unit with installed fault condition

SAFETY
Wear your safety goggles at all times. When working with electrical circuits, be sure to use insulated tools to avoid electrical shock.

PROCEDURE
1. Wearing safety goggles, connect the gauge set to the unit.
2. Install the thermometers at the following locations to measure:
   a. Condenser entering air or liquid temperature.
   b. Condenser leaving air or liquid temperature.
   c. Evaporator entering air or liquid temperature.
   d. Evaporator leaving air or liquid temperature.
   e. Suction vapor temperature at compressor.
   f. Liquid temperature leaving the condenser.
3. Connect the ammeter on one of the power leads to the compressor motor.
4. Start the unit and record the information as outlined in text book.
5. Review the information found and compare to problem listings in text book for electrical problems and/or for refrigeration operation.
6. Determine the approximate cause of problem and determine the listing number.
7. Proceed to Chapter R49, Trouble-Shooting - Refrigeration- Electrical, for possible solution.

Repeat this procedure for each problem set up by instructor.

Laboratory Worksheet - # R40
Student___________________________
Instructor __________________________
Date _____________ Grade__________

PREVENTIVE MAINTENANCE

EXPECTED COMPETENCY:

To demonstrate the ability to perform preventive maintenance on heating and air conditioning equipment

STUDY MATERIAL

Chapter PM53

EQUIPMENT

1. Safety goggles
2. Air conditioning split system
3. Heating system-either gas, oil, or electric heat
4. Refrigeration gauges
5. 2 electronic thermometers
6. VOM with amp clamp
7. Student tool box

SAFETY

Be cautious when working around moving parts when performing maintenance. It is best to shut down the equipment, lockout, and then tagout prior to working on any equipment.

PROCEDURE

1. Locate the power supply to the air conditioner and then lockout-tagout.
2. Locate and perform inspections and maintenance on the following components, indicate condition of the equipment below:
   Indoor air filter _____________
   Indoor fan motor _____________
   Outdoor fan motor _____________
   Evaporator coil _____________
   Condenser coil _____________
   Electrical connections_____________
   Thermostat _____________
3. Re-install all covers on the equipment, remove lockout-tagout safeties, and then turn on equipment and test for the following:
   Evaporator coil temperature difference _____________
   Outdoor air coil temperature difference _____________
   Suction pressure _____________
   Discharge pressure _____________

Compressor amps ____________
Suction superheat ____________

4. Locate the power supply to the heating unit and then lockout-tagout.
5. Locate and perform inspections and maintenance on the following components, indicate condition of the equipment below:
   Indoor air filter ____________
   Indoor fan motor ____________
   Electrical connections ____________
   Ignition controls ____________
   Thermostat ____________

6. Re-install all covers on the equipment, remove lockout-tagout safeties, and then turn on equipment and test for the following:
   Heating surface temperature difference ____________
   Blower amps ____________

QUESTIONS

1. How many times during a year should preventive maintenance be performed on air conditioning and heating equipment?

2. What should you do when you find coil air temperatures are exceptionally higher than normal?

3. How is the thermostat calibrated?

4. What is the indication that the evaporator coil or condensing coil is getting dirty?

5. What is the procedure if you find a small pool of oil and dirt around a refrigeration gauge fitting on the condensing unit?

Laboratory Worksheet - # E1
Student______________________
Instructor ______________________
Date ___________ Grade_________

BASIC ELECTRICAL STATIC CHARGE

EXPECTED COMPETENCY:
To become acquainted with the electrical forces between charged bodies

STUDY MATERIAL
Chapter E19

EQUIPMENT
1. Glass rod
2. Hard rubber rod
3. Pieces of silk and animal fur
4. Pith balls
5. Light thread

PROCEDURE
1. Tie the pith balls to the glass rod.
2. Rub the rod with the piece of silk and observe the action of the balls. Record the information on the chart below.
3. Repeat steps 1 and 2 using the hard rubber rod.

Step 1		Step 2	
Type of rod used	______________	Type of rod used	______________
Material rubbed	______________	Material rubbed	______________
Force indicated	______________	Force indicated	______________

4. Repeat steps 3 and 4 using the animal fur.

Step 3		Step 4	
Type of rod used	______________	Type of rod used	______________
Material rubbed	______________	Material rubbed	______________
Force indicated	______________	Force indicated	______________

QUESTIONS
1. Why do the balls in steps 1 and 2 separate?

2. Why do the balls in steps 3 and 4 try to touch?

Laboratory Worksheet - # E2
Student______________________________
Instructor ____________________________
Date ______________ Grade__________

BASIC MAGNETISM

EXPECTED COMPETENCY:
To become acquainted with magnetic action and lines of force

STUDY MATERIAL
Chapter E19

EQUIPMENT
1. Two bar magnets
2. Flat piece of Lucite on corner posts
3. Quantity of iron filings
4. Light thread
5. Glass rod

PROCEDURE
1. Suspend each magnet, in turn, by means of the light thread around the balance center of the magnet. Magnet should balance on the thread between the magnet and the rod.
2. Allow the magnet to seek the earth compass direction.
3. Mark with an "S" the end of the magnet that turns to the north.
4. Mark the other end with an "N".
5. Do the same with the second magnet.
6. Place one of the magnets under the Lucite and sprinkle the iron filings on the Lucite.
7. Draw the paths that the filings assume.
8. Slide the second magnet under the Lucite to bring the two "N" ends together.
9. Observe the action of the filings.
10. Repeat this action using the two "S" ends of the magnets.
11. Repeat the action by placing the unlike ends together, "N" to "S".
12. Observe the action of the filings.

QUESTIONS
1. When two like ends (poles) of the magnets were placed together, the magnets wanted to move apart. Why?

2. When unlike ends of the magnets were brought together, they rapidly moved together. Why?

3. The basic rule of magnetism is, therefore, like poles __________and unlike poles __________.

Laboratory Worksheet - # E3
Student________________________
Instructor ______________________
Date ____________ Grade_________

ELECTROMOTIVE FORCE

EXPECTED COMPETENCY:
To become acquainted with the production of electrical energy by the action of a magnetic field

STUDY MATERIAL
Chapter E19

EQUIPMENT
1. A spool of fine wire
2. Cardboard tube (light material)
3. Magnet that will enter tube
4. Voltmeter

PROCEDURE
1. Wind the wire around the cardboard tube at least 20 turns. Hold in place by covering with clear plastic tape. Allow a minimum of 12 ft. leads.
2. Remove the insulation from the ends of the wire and connect the leads to the voltmeter.
3. Slide the magnet in and out of the tube.
4. Observe the action of the voltmeter as the magnet enters, stops and leaves the tube.

QUESTIONS
1. The voltmeter indicates the flow of electrical energy as the magnet enters and leaves the tube. Why?
2. Is there any indication of current flow when the magnet is stationary?
3. Why does the needle indicate a different direction of current flow when the magnet is entering and leaving the coil?

Laboratory Worksheet - # E4

Student______________________________

Instructor ____________________________

Date _____________ Grade__________

CHEMICAL ACTION FORCE

EXPECTED COMPETENCY:

To demonstrate the conversion of chemical action to electromotive force

STUDY MATERIAL

Chapter E19

EQUIPMENT

1. One large lemon
2. Wedge-shaped pieces of copper and zinc, pointed at one end and with a 12" piece of wire fastened to the other end of each piece.
3. Voltmeter

PROCEDURE

1. Clean the insulation from the ends and fasten the wire leads to the voltmeter.
2. Insert the metal strips through the lemon so they are parallel and vertical to the core of the lemon.
3. Observe the action of the voltmeter.
4. Allow the action to continue until it tapers off and stops.
5. Remove the metal strips for inspection.

QUESTIONS

1. What change took place on the surface of the copper strip?

2. What change took place on the surface of the zinc strip?

3. Would it be possible to recharge the lemon battery?

Laboratory Worksheet - # E5
Student___________________________
Instructor ___________________________
Date _____________ Grade__________

HEAT ACTION FORCE

EXPECTED COMPETENCY:
To demonstrate the process of converting heat energy to electrical energy

STUDY MATERIAL
Chapter E19

EQUIPMENT
1. Standard gas unit thermocouple
2. Milli-voltmeter
3. Electrical leads with clips
4. Two Bunsen burners

SAFETY
Be cautious when working around the Bunsen burners to avoid burning your skin.

PROCEDURE
1. Connect the electrical leads to the terminals of the voltmeter.
2. Connect one of the lead clips to the connector stub of the thermocouple.
3. Connect the other lead to the copper shell of the thermocouple.
4. Apply heat to the thermocouple tip and watch the action of the needle on the milli-voltmeter. If the needle moves off the scale in the wrong direction, reverse the lead clips.
5. With the lead clips connected correctly, heat the tip of the thermocouple using one of the Bunsen burners until maximum output is produced. Record the milli-volt output.
6. While continuing to heat the tip of the thermocouple, use the second Bunsen burner to heat the body of the thermocouple at the mounting ring. Record the action of the milli-voltmeter.

QUESTIONS
1. Is the output of a thermocouple A.C. or D.C. milli-volts?

2. What happens when the body of the thermocouple is also heated?

3. The output of a thermocouple depends upon (name one)
   a. The temperature of the tip.
   b. The temperature of the body.
   c. The length of the tip.
   d. The temperature difference between the body and the tip.

Laboratory Worksheet - # E6

Student_________________________

Instructor _______________________

Date ____________ Grade_________

ELECTRICAL CIRCUITS

EXPECTED COMPETENCY:

To become acquainted with the different types of electrical circuits and the variations in current flow with various resistances

STUDY MATERIAL

Chapter E20

EQUIPMENT

1. Four dry-cell batteries
2. Light bulb test board
3. 9-volt DC bulb
4. 120-volt light cord
5. AC/DC voltmeter and ammeter
6. Connection wires
7. Student tool box
8. Light bulbs (40, 60, 100 watt)

PROCEDURE

1. Connect the 9-volt bulb to the following combinations of dry cell batteries. Record the voltage across the bulb and current flow through the bulb.

	Voltage	Current
1 battery	_______	_______
2 batteries in series	_______	_______
2 batteries in parallel	_______	_______
3 batteries in series	_______	_______
3 batteries in parallel	_______	_______
4 batteries in series	_______	_______
4 batteries in parallel	_______	_______

2. Remove the batteries and wires.
3. Connect the 120-volt light cord to the light bulb test board.
4. Record the voltage across each bulb and the current flow through each bulb as well as the total current flow through the test cord for each of the following bulb combinations.

Combination	Voltage across each bulb	Current flow through each bulb	Total current
40-watt bulb	__________	__________	_______
40W & 60W in series	______, ______	______, ______	_______
40W & 60W in parallel	______, ______	______, ______	_______
40W, 60W & 100W in series	______, ______ ______	______, ______ ______	_______
40W, 60W & 100W in parallel	______, ______ _______	______, ______ _______	_______

QUESTIONS

1. When two batteries were connected in series, what happened to the voltage and current compared to the single battery in series?

2. When two batteries were in parallel, what change took place as compared to the single battery?

3. When three batteries were connected in series, what happened to the voltage and current compared to the single battery?

4. When three batteries were connected in parallel, what happened to the voltage and current compared to the single battery?

5. When four batteries were connected in series, what happened to the voltage and current as compared to the single battery in series?

6. When four batteries were connected in parallel, what happened to the voltage and current as compared to the single battery in series?

7. What is the voltage across and the current through the 40-watt bulb?

8. With the 40-watt and 60-watt bulbs in series, what is the voltage across each bulb, the current through each bulb, and the total current?

9. With the 40-watt and 60-watt bulbs in parallel, what is the voltage across each bulb, the current through each bulb, and the total current ?

10. With the 40-watt, 60-watt and 100-watt bulbs in series, what is the voltage across each bulb, the current through each bulb, and the total current?

11. With the 40-watt, 60-watt and 100-watt bulbs in parallel, what is the voltage across each bulb, the current through each bulb, and the total current?

12. In multiple loads in series, the __________divides among the loads and the total current is ______than in the same loads in parallel.

Laboratory Worksheet - # E7
Student______________________
Instructor ______________________
Date ____________ Grade_________

ELECTRICAL CALCULATIONS (OHM'S LAW)

EXPECTED COMPETENCY:
To acquire knowledge and skills in the calculation of current and resistance in series and parallel circuits

STUDY MATERIAL
Chapter E19

EQUIPMENT
1. Light bulb test board
2. Light bulbs (40, 60, 100 and 150 watts)
3. A.C. voltmeter and ammeter
4. Ohm meter

PROCEDURE
1. With the light bulbs set in the test board as specified below, calculate the current and resistance in each arrangement. Measured voltage is ?

Light bulbs	Calculations		Measured	
in series	Current	Resistance	Current	Resistance
40, 60 & 150 watts	______	______	______	______
40, 100 & 150 watts	______	______	______	______
60 & 100 watts	______	______	______	______
60 & 150 watts	______	______	______	______
60, 100 & 150 watts	______	______	______	______
100 & 150 watts	______	______	______	______

Light bulbs	Calculations		Measured	
in parallel	Current	Resistance	Current	Resistance
40, 60 & 150 watts	______	______	______	______
40, 100 & 150 watts	______	______	______	______
60 & 100 watts	______	______	______	______
60 & 150 watts	______	______	______	______
60, 100 & 150 watts	______	______	______	______
100 & 150 watts	______	______	______	______

QUESTIONS

1. The total current through a series circuit is determined by (Which statements are correct?):
   a. The applied voltage and each load resistance.
   b. The voltage across each load and the resistance of each load.
   c. The voltage across the entire circuit and the total resistance of all loads.

2. The voltage across each load in a series circuit will:
   a. Divide equally across the various loads.
   b. Divide in proportion to the resistance of each load.
   c. All be the same as the power source voltage.

3. The total current through a parallel circuit is determined by (Which statements are correct?):
   a. The applied voltage and each load resistance.
   b. The voltage across each load and the resistance of each load.
   c. The voltage across the entire circuits and the total resistance of all the loads.

4. The voltage across each load in a parallel circuit will:
   a. Divide equally across the various loads.
   b. Divide in proportion to the resistance of each load.
   c. All be the same as the power source voltage.

Laboratory Worksheet - # E8
Student________________________
Instructor ______________________
Date ___________ Grade_________

INSPECT, CONNECT, AND OPERATE TRANSFORMERS

EXPECTED COMPETENCY:
To become acquainted with the action and identification of transformers

STUDY MATERIAL
Chapter E20

EQUIPMENT
1. 120/24V transformer
2. 240/24V transformer
3. 120V oil burner ignition transformer
4. Voltmeter, clamp type ammeter and ohm meter
5. Light board

PROCEDURE
1. Determine the primary and secondary side of each of the three transformers.
2. Connect the primary side of the 120/24V transformer to 120V power supply. Measure the output voltage.
3. Connect the 24-volt side of the 120/24V transformer to the bulb sockets on the light board.
4. Measure the output voltage and amperage of the transformer with the following bulb combinations:

Bulb combinations	Output voltage	Output amperage
25-watt	_______	_______
25 & 40 watt	_______	_______
25, 40 & 60 watt	_______	_______
25, 40, 60 & 100 watt	_______	_______

5. Connect the 24-volt side of the 120/24V and 240/24V transformers.
6. Connect the 120/24V transformer primary to 120V power supply.
7. Measure the voltage on the primary side of the 240/24V transformer.

QUESTIONS

1. Which winding has the higher resistance in a step-down transformer?

2. Which winding has the higher resistance in the oil burner ignition transformer? Why?

3. As the load is increased on the light board, what happens to the output voltage and amperage of the transformer?

4. With the two transformers connected together, what is the output voltage of the second transformer? _________volts.

5. If a 240V/24V transformer had the 24-volt side connected to 120 volts, for the few seconds the transformer would last, what would be the output voltage? __________volts.

6. The primary and secondary windings in a transformer are:
   a. Wired in parallel
   b. Wired in series
   c. Not connected electrically
   d. Connected with a solenoid switching device

7. Describe why the voltage is different between the input and the output of a transformer?

8. A typical control transformer is a (step-up or step-down) transformer.

Laboratory Worksheet - # E9

Student____________________________

Instructor ___________________________

Date _____________ Grade__________

ELECTRICAL LOADS

EXPECTED COMPETENCY:

To become acquainted with the types of loads encountered in electrical control circuits

STUDY MATERIAL

Chapter E20

EQUIPMENT

1. Container of various types of loads, both magnetic and thermal as well as primary and parasitic

PROCEDURE

1. Examine each part in turn and list if it is a thermal or magnetic type as well as primary or parasitic.

Part	Thermal or magnetic	Primary or parasitic

QUESTIONS

1. Identify a magnetic type load.

2. Identify a thermal type load.

3. Identify a primary type load.

4. Identify a parasitic type load.

Laboratory Worksheet - # E10
Student______________________________
Instructor ____________________________
Date ____________ Grade__________

ELECTRICAL CONTROLS

EXPECTED COMPETENCY:
To become acquainted with the various types of controls used in the refrigeration field

STUDY MATERIAL
Chapter E20

EQUIPMENT
1. Box containing controls of all types for identification

PROCEDURE
Examine each control and record its characteristics.

	Control name	Actuated by	Capacity amperes	Primary or secondary
1.	________________	__________	________	________
2.	________________	__________	________	________
3.	________________	__________	________	________
4.	________________	__________	________	________
5.	________________	__________	________	________
6.	________________	__________	________	________
7.	________________	__________	________	________
8.	________________	__________	________	________
9.	________________	__________	________	________
10.	________________	__________	________	________

QUESTIONS
1. Define a primary type control. Give an example.

2. Define a secondary type control. Give an example.

3. Describe the difference between a relay, contactor and a motor starter.

Laboratory Worksheet - # E11

Student____________________________

Instructor ___________________________

Date ____________ Grade_________

CAPACITORS

EXPECTED COMPETENCY:

To become acquainted with the types of capacitors used in the refrigeration field, their construction and ratings

STUDY MATERIAL

Chapter E21

EQUIPMENT

1. Cutaway models of start and run capacitor
2. Several run capacitors of different sizes and voltages
3. Several start capacitors of different sizes and voltages

SAFETY

Be cautious when working around old capacitors that might have PCB toxins and oils inside. Be sure to read the label and, if unsure, consult with your instructor to verify type.

PROCEDURE

1. List the capacity and working voltage of the capacitor types.

Type	Capacity	Working voltage
__________________	__________	_______________
__________________	__________	_______________
__________________	__________	_______________
__________________	__________	_______________
__________________	__________	_______________
__________________	__________	_______________
__________________	__________	_______________
__________________	__________	_______________

QUESTIONS

1. What would be the total capacity and working voltage if three capacitors of 40 mfd, 20 mfd and 10 mfd were connected in parallel?

2. Can a capacitor of 40 mfd/370 W.V. and a capacitor of 20 mfd/250 W.V. be connected in parallel across a motor application calling for 320 working voltage?

3. Using the following formulae:

$$\text{Total mfd} = \frac{\text{Cap.1 mfd x Cap.2 mfd.}}{\text{Cap.1 mfd + Cap.2 mfd.}}$$

a. What would be the total capacity if a 50 mfd capacitor and a 40 mfd capacitor were connected in series?

4. If the 50 mfd capacitor had a working voltage of 240 volts and the 40 mfd capacitor had a working voltage of 120 volts, what would be the working voltage of the two capacitors connected in series?

5. How do the insides of a start capacitor differ from a run capacitor?

6. What is the purpose of the resistor when used on capacitors?

7. How is an open circuit inside a capacitor detected?

8. How is a short circuit inside a capacitor detected?

9. What is the purpose of the run capacitor?

10. Describe the differences in construction between the start and run capacitors.

Laboratory Worksheet - # E12

Student________________________

Instructor ________________________

Date ____________ Grade_________

DISASSEMBLE AND REASSEMBLE A CAPACITOR START TYPE MOTOR

EXPECTED COMPETENCY:

To become acquainted with the construction and minor maintenance of a capacitor start type motor

STUDY MATERIAL

Chapter E21

EQUIPMENT

1. Capacitor type motor
2. Clamp ammeter
3. Student tool kit

SAFETY

Avoid touching wires together when measuring amps. This could cause electrical shock and injury.

PROCEDURE

1. Locate and make marks on the end bells and stator so the motor can be re-assembled correctly.
2. Remove the assembly bolts and place the nuts back on them to prevent loss.
3. Remove the end bells and rotor.
4. Examine the rotor for evidence of heating and/or an open circuit.
5. Note the connections to the starting switch and to the capacitor. Trace the windings and draw a diagram showing windings, starting switch and capacitor.
6. Reassemble the motor.
7. With the motor running, use a pencil to mark the shaft at the pulley end, even with the end bell bearing projection. Stop the motor and move the rotor end-ways by hand. There should be a slight amount of end play in each direction from the spot marked.
8. If the end play is incorrect, disassemble the motor and correct the condition by rearranging shims or adding some, if necessary.
9. Take ampere reading of the motor while running unloaded.
10. Record: nameplate ampere rating ________, horsepower rating________, speed _______.

QUESTIONS

1. What is the theoretical fastest speed at which a two-pole motor can operate?

2. What is the full-load speed of a two-pole motor?

3. What is the full-load speed of a four-pole motor?

4. The capacitor is connected in series with which one of the motor windings?

5. What change is necessary to make a motor start and operate in the reverse direction?

6. Name the methods of controlling the operation of the starting circuit of a capacitor start motor.

7. Define a permanent split-capacitor type motor.

8. A capacitor start - capacitor run motor uses a capacitor to start the motor and another capacitor to run the motor. True or False?

9. Describe how a capacitor functions in a starting circuit and in a running circuit.

Laboratory Worksheet - # E13

Student___________________________

Instructor __________________________

Date ____________ Grade_________

ELECTRICAL TESTING DEVICES

EXPECTED COMPETENCY:

To become familiar with the electrical instruments used to service a refrigeration system

STUDY MATERIAL

Chapters R6 and E20

EQUIPMENT

1. Safety goggles
2. Operating refrigeration system. May be a remote condensing unit walk-in cooler
3. A clamp type ammeter/voltmeter. Voltage range must be as high as 600 volts
4. VOM/ohm meter - must be a multi-scale with low scale capable of measuring 0-10 ohms accurately
5. Megger insulation tester
6. Student tool set

SAFETY

Be cautious when working around the Megger insulation tester. This device generates high voltage to test insulation break-down, and can be potentially dangerous. Always use insulated tools and test equipment when working with any high-voltage equipment.

PROCEDURE

1. Safety goggles must be worn at all times around electrical equipment.
2. Remove the access panels from the electrical control panel and compressor terminal box.
3. Using the voltmeter, measure and record the line voltage with the unit off.
4. Clamp the ammeter around one of the power leads to the unit.
5. While watching the amperage draw, start the unit. Amperage should indicate a high draw and fall immediately to run amperage. If amperage stays high, disconnect the power.
6. If unit does not start, trace wiring and check condition of each part.
7. Disconnect leads to compressor to operate condenser fan only.
8. With ammeter clamped to one lead to fan motor, apply power and check amperage draw. If it remains high, replace the fan motor.
9. Check the compressor circuit component parts.
   a. Overload - use ohm meter to check continuity.
   b. Run capacitor - use ohm meter to check condition.
   c. Start capacitor - use ohm meter to check condition.
10. Use ohm meter to check winding resistance of the compressor motor.
11. Use Megger to check the electrical lead insulation between each lead and ground (the compressor shell).

QUESTIONS

1. Why must the electrical amperage draw drop within two seconds?

2. If a current type relay is used to start the compressor, will the ohm meter reading across the contacts be infinite?

3. When measuring the resistance of the motor windings, which winding has the greater resistance, the run or the start winding?

4. Electrical insulation of the motor should measure more than:
   a. 10,000 ohms
   b. 50,000 ohms
   c. 100,000 ohms
   d. 1,000,000 ohms

5. When checking the capacitors, if the ohm meter needle remains at infinity on the scale, what is the condition of the capacitor?

6. If the needle travels to and stays at the "0" ohm reading, what is the problem with the capacitor?

7. Why must the power be off when an ohm meter is used?

8. Can a capacitor be checked without removing the connectors? Why or why not?

9. What are the two main categories of motor problems?

10. How is a stuck hermetic compressor repaired?

11. What three electrical conditions must be met to have current flow?

12. What two electrical test instruments are most often used for motor problems?

13. Where is a convenient place to check a motor for an electrical ground?

Laboratory Worksheet - #E14
Student________________________
Instructor ________________________
Date ____________ Grade_________

ELECTRICAL RELAYS AND CONTACTORS

EXPECTED COMPETENCY:

To become familiar with the different types of relay and contactor arrangements used in electrical controls

STUDY MATERIAL

Chapters E20

EQUIPMENT

1. Safety goggles
2. Student tool set
3. Box of controls, relays, switches, etc.

PROCEDURE

Inspect each individual control. Record the method of actuation and the contact arrangement.

Item	Actuation method	Number of contacts	N.O. or N.C.

QUESTIONS

1. Define the following switch arrangements.
   a. S.P.S.T. c. D.P.S.T.
   b. S.P.D.T. d. D.P.D.T.

2. A compressor contactor will usually have what contact arrangement provided by the manufacturer?

3. A fan control will have what contact arrangement?

4. The contact in a fan relay is normally open or normally closed?

5. A cooling thermostat is shown on a wiring diagram as normally open or normally closed?

Laboratory Worksheet - # E15
Student______________________
Instructor ______________________
Date ____________ Grade__________

ELECTRONIC IGNITION SYSTEMS

EXPECTED COMPETENCY:
To become familiar with the electronic ignition systems used in gas heating units

STUDY MATERIAL
Chapter H36

EQUIPMENT
1. Safety goggles
2. Gas-fired heating unit equipped with electronic intermittent pilot ignition system
3. Gas-fired heating unit equipped with direct spark ignition system
4. Student tool set

SAFETY
Be cautious to avoid electrical shock when working around the ignition transformer. It generates several thousand volts during the ignition cycle.

PROCEDURE
1. Wire the electronic intermittent pilot ignition system according to the manufacturer's wiring diagram.
2. Operate the unit through a complete on/off cycle.
3. Record the steps that take place in the operating cycle.
4. Wire the direct spark ignition system according to the manufacturer's wiring diagram.
5. Operate the unit through a complete on/off cycle.
6. Record the steps that take place in the operating cycle.

QUESTIONS
1. What is the major difference between the two systems?

2. What is the usual voltage produced by the system across the igniter leads?

3. How does low voltage affect the spark ignition system?

Laboratory Worksheet - # E16
Student________________________
Instructor ______________________
Date ___________ Grade_________

DIRECT DIGITAL CONTROLS

EXPECTED COMPETENCY:
To demonstrate the ability to work with direct digital controls and control accessories for controlling HVAC equipment

STUDY MATERIAL
Chapter A31

EQUIPMENT
1. Safety goggles
2. VOM meter
3. DDC control panel
4. Computer or field interface monitor
5. DDC equipment startup and service literature
6. Control devices, including valves, dampers, relays, etc.
7. Control sensors, including RTD's, thermistors, transducers, etc.
8. Student tool box

SAFETY
Be careful to avoid static electricity when working around electronic controls. Work with grounded equipment and instruments. This will help prevent electrical damage to controller boards and microprocessor memories.

PROCEDURE
1. Familiarize yourself with the DDC controller, mother board, power supply, and field termination board.
2. Draw on a sheet of paper the control interface accessories, including all of the controlled devices and sensors.
3. Identify the control loop, open or closed, process variable, and control action.
4. Identify all digital outputs, digital inputs, analog outputs, analog inputs.
5. Review the startup and service literature associated with the equipment.
6. Power up the DDC control panel, log on to the system, and monitor existing control conditions.
7. Turn on the digital outputs.
8. Modulate all of the analog outputs from fully open to fully closed.
9. Release priority condition to computer control.
10. Adjust set points to allow control system to adjust control process.

11. Record the following:

	Digital outputs	Digital inputs
	____________	____________
	____________	____________
	Analog outputs	Analog inputs
	____________	____________
	____________	____________

QUESTIONS

1. What type of control loop were you working with, open or closed?

2. What was the process variable you were controlling?

3. Was the control action direct or reverse-acting?

4. Was the DDC panel an intelligent stand-alone device, or did it need the host computer to operate?

5. When you adjusted the set point, did the DDC panel respond immediately or was it relatively slow to adjust the output signal? Explain.

6. Draw a simple schematic of the DDC control system below.

Laboratory Worksheet - # A1

Student ____________________

Instructor ____________________

Date __________ Grade __________

PSYCHROMETRICS

EXPECTED COMPETENCY:

To become acquainted with the properties of air and the use of the sling psychrometer and psychrometric chart

STUDY MATERIAL

Chapter A28

EQUIPMENT

1. Use the laboratory air distribution system with fresh air supply.
2. Sling psychrometer
3. Psychrometric chart

PROCEDURE

1. Measure the wet and dry bulb temperatures of the fresh air, return air, mixture and supply air.
2. Determine the percent of relative humidity and the dew point.

Air sample	Dry bulb	Wet bulb	R.H. %	Dew point
Return air	________	________	________	________
Fresh air	________	________	________	________
Supply air	________	________	________	________
Mixture	________	________	________	________

3. Determine the enthalpy and humidity ratio of each air sample.

Air sample	Enthalpy	Humidity ratio
Return air	__________	__________
Fresh air	__________	__________
Supply air	__________	__________
Mixture	__________	__________

4. Reduce the amount of fresh air into the building. After five minutes of operation, repeat steps 1 through 3.
5. Increase the amount of fresh air to a higher level than the original amount.
6. Repeat steps 1 through 3.
7. Close off all the fresh air supply.
8. Repeat steps 1 through 3.
9. Return the fresh air supply to the original amount.

QUESTIONS

1. What effect did the cutting off the fresh air supply have on the conditions of the air entering the air handling unit?

2. What effect did increasing the fresh air supply have on the conditions of the mixture? Why?

3. What effect did increasing the fresh air supply have on the conditions of the supply air?

4. What causes the thermometer with the wet sock to read lower than the thermometer with the dry bulb?

5. Why must the thermometer be swung in order to obtain the reading?

6. What does it mean when it is said that air is 70% R.H.?

Laboratory Worksheet - # A2
Student_______________________
Instructor _____________________
Date ____________ Grade__________

AIR FLOW PRINCIPLES

EXPECTED COMPETENCY:
To become acquainted with the measurement of air quantity and flow rate in air distribution systems

STUDY MATERIAL
Chapter H40

EQUIPMENT
1. Laboratory air condition system
2. Inclined manometer with sampling tubes
3. Pitot tube
4. Anemometer
5. A balameter

PROCEDURE
1. Use the inclined manometer to measure both supply and return static pressures at the air handling unit.
2. Use inclined manometer and Pitot tube assembly to determine air velocity in the duct.
3. Calculate air quantity through the duct.
4. Measure air supply value from each outlet with anemometer and balameter.
5. Determine total air quantity through the duct.
6. Check calculations against Pitot tube measurement and calculations.
7. Measure and record the static pressure in the duct down stream from each outlet or branch run.

QUESTIONS
1. Which method of measuring air quantity is the most accurate, by velocity times the area or by total air from outlets?

2. What is the definition of E.S.P. (external static pressure)?

3. If the friction loss through a 100 ft. length of 6" x 22" supply duct is 0.25" W.C. what is the C.F.M. of air through the duct?

4. What would be the static pressure drop through an 8" x 16" duct if 450 C.F.M. were forced through the duct? Static pressure drop =__________.

Laboratory Worksheet - # A3

Student______________________

Instructor ______________________

Date ____________ Grade_________

AIR MIXTURES

EXPECTED COMPETENCY:

To become acquainted with the handling of air mixtures

STUDY MATERIAL

Chapter A32

EQUIPMENT

1. Use the laboratory air distribution system with fresh air supply.
2. Sling psychrometer
3. Psychrometric chart
4. Inclined manometer with Pitot tube
5. A balameter

PROCEDURE

1. Measure the total supply air quantity using both the inclined manometer and Pitot tube combination and the balameter.
   a. Instrument C.F.M.
   b. Inclined manometer
   c. Balameter
2. Measure the dry and wet bulb temperatures of the return air, fresh air and mixture entering the air handling unit.
3. Calculate the amount of return air and fresh air in the mixture.
4. Calculate the total heat content of the return air, fresh air and air mixture.
5. Calculate the heat content of the supply air.
6. Calculate the total heat removed or added by the air conditioning unit.
7. Reduce the amount of fresh air and repeat steps 1 through 6.
8. Increase the amount of fresh air over the original amount and repeat steps 1 through 6.

RETURN FRESH AIR SETTING TO ORIGINAL AMOUNT.

QUESTIONS

1. What effect does the increase in fresh air have on the amount of heat removed or added by the air conditioning unit?

2. What effect does the decrease in fresh air have on the sensible/ total ratio of the air conditioning unit?

3. What effect does the increase in fresh air have on the heating or cooling capacity of the air conditioning unit?

4. What is the formula used in the field to calculate mixed air percentages?

Laboratory Worksheet - # A4
Student__________________________
Instructor ________________________
Date ____________ Grade_________

AIR TEMPERATURE CONTROLS

EXPECTED COMPETENCY:
To become acquainted with the various types of room temperature controls, both single season and year round operation

STUDY MATERIAL
Chapter A31

EQUIPMENT
1. Box of various types of room temperature controls
   a. Snap action
   b. Mercury tube
   c. Heating sub-base
   d. Cooling sub-base
   e. Year round sub-base
   f. Two-stage heating
   g. Two-stage cooling

PROCEDURE
1. Examine the control and record the following:
   a. The methods of making and breaking the control circuit.
   b. The methods of changing the desired temperature level to be maintained.
2. Trace the circuit through each circuit of the control-heating and/or cooling.

QUESTIONS
1. What is the main disadvantage of open snap-action contacts?

2. What is the main disadvantage of mercury tube contacts?

3. What is the minimum temperature range between the two steps of heating or cooling?

4. In an automatic change-over type year round thermostat, what is the minimum temperature range between the heating and cooling temperatures that can be maintained?

5. Which type of thermometer would respond the fastest, a mercury bulb or an electronic thermistor?

6. How does a bimetal sensing element make and break an electrical circuit?

7. Describe a good location for a space temperature thermostat?

Laboratory Worksheet - # A5

Student__________________________

Instructor __________________________

Date _____________ Grade__________

HUMIDISTATS

EXPECTED COMPETENCY:
To become acquainted with the construction and operation of humidity controls

STUDY MATERIAL
Chapter H35

EQUIPMENT

1. Humidistat using human hair
2. Humidistat using nylon strip
3. Humidistat using coated metal strip coil

PROCEDURE
Examine each type of control and explain the difference in operation.

QUESTIONS

1. A rise in humidity causes the human hair to expand or contract?

2. A drop in humidity causes the Teflon strip to expand or contract?

3. What effect does a change in humidity setting have on the power element in the control?

4. What action does a change in humidity produce on the coated metal strip coil?

Laboratory Worksheet - # A6
Student_______________________
Instructor _____________________
Date ___________ Grade_________

SENSING DEVICES

EXPECTED COMPETENCY:
To become acquainted with types of sensing devices used for operation and/or protection of heating and air conditioning systems

STUDY MATERIAL
Chapter H35

EQUIPMENT
A collection of the various types of operating controls, limit controls, and safety controls employing the following types of operators.
1. Bi-metallic element
2. Bellows and diaphragms
3. Bourdon tube
4. Rod and tube

PROCEDURE
Examine and write an explanation of how each control operates:
1. Bi-metallic element

2. Bellows and diaphragms

3. Bourdon tube

4. Rod and tube

QUESTIONS
1. A temperature control would use what types of operators?

2. A pressure control would use what types of operators?

3. A limit control using a thermostatic element would (open or close) the control circuit on a temperature rise.

4. The most common type of operator in a low pressure control is the ____________ type.

5. Rod and tube operators are normally only used in pilot controls. Why?

Laboratory Worksheet - # A7

Student_______________________

Instructor _______________________

Date ____________ Grade_________

GAS CONTROLS

EXPECTED COMPETENCY:
To become acquainted with the controls used on gas-fired heating equipment

STUDY MATERIAL
Chapters H35 and H36

EQUIPMENT
Samples of:
a. Gas pressure regulators
b. Pilot safety switches
c. Solenoid valves
d. Combination gas valves

PROCEDURE
1. Disassemble and re-assemble each control to determine the sequence of operation.
2. Record the type of valve and explanation of the operation for future reference.

If there are any questions as to the operating steps, consult your instructor.

QUESTIONS
1. What is the typical gas manifold pressure for a natural gas furnace?
2. What is the typical pressure for an LP furnace manifold?
3. Why does natural gas require a 100% shut off?
4. Why is it necessary to have a pilot safety shutoff on a gas furnace?
5. Describe how a thermocouple works.
6. What component reduces the main pressure for a typical gas furnace?

Laboratory Worksheet - # A8
Student ____________________
Instructor ____________________
Date ____________ Grade ________

ELECTRONIC IGNITION SYSTEMS

EXPECTED COMPETENCY:
To become acquainted with the construction and operation of the different types of electronic ignition systems

STUDY MATERIAL
Chapter H36 and H36

EQUIPMENT
1. Control board containing burner set-up, using intermittent pilot burner
2. Control board containing burner set-up, using direct spark ignition

SAFETY
Wear safety goggles, don't over-pressurize the gas pressure, (3½" W.C. for natural gas and 11" W.C. for liquid propane) and provide adequate ventilation during the exercise.

PROCEDURE
1. Connect each board in turn to a gas fuel supply and controls.
2. Operate each burner set-up through the complete operating cycle.
3. Record the steps each control system takes to become operative.
4. Shut off the fuel supply to each board set-up and record the step-by-step action each control set-up takes.
5. Reverse the polarity of the power supply to the systems and observe the action.

QUESTIONS
1. Which system responds faster upon demand from the room thermostat?
2. Which system responds faster upon fuel supply failure?
3. What effect does the power supply reversed polarity have on the system's operation?
4. Explain the operation of each system after a momentary shut-off of power.

Laboratory Worksheet - # A9
Student______________________________
Instructor ____________________________
Date ______________ Grade__________

OIL BURNER CONTROLS

EXPECTED COMPETENCY:
To become acquainted with the operation of oil burner controls

STUDY MATERIAL
Chapters H35 and H36

EQUIPMENT
1. Stack-mounted Protecto-relay
2. Cad cell Protecto-relay
3. Delayed action oil valve

PROCEDURE
1. Examine each control to determine its sequence of operation.
2. Record the sequence of operation of each type of control for future reference .
   a. Stack-mounted Protecto-relay

   b. Cad cell Protecto-relay

   c. Delayed action oil valve

QUESTIONS
1. What method must be used to put a stack-mounted relay back in sequence?

2. What would be the major cause of operating failure of a stack-mounted relay?

3. What would be the major cause of failure of a cad cell relay?

4. What is the primary purpose of a delayed action oil valve?

5. What is the function of the stack switch on an oil furnace?

6. Why is the cad cell used on an oil furnace and not on a gas furnace?

7. What would be the symptoms of a dirty cad cell?

8. A stack switch has what kind of sensing element?

9. If the burner does not fire in the prescribed time, how do you restart it?

Laboratory Worksheet - # A10

Student_______________________

Instructor _______________________

Date ____________ Grade_________

HYDRONIC HEAT

EXPECTED COMPETENCY:

To demonstrate the ability to start up, operate, and troubleshoot a residential or light commercial hydronic water heating system

STUDY MATERIAL

Chapter H39

EQUIPMENT

1. Safety goggles
2. Hydronic furnace system
3. Circulation pump(s) with unit heaters with air purge valves
4. Fuel combustion efficiency testing instrument
5. Four electronic thermometers
6. VOM with amp clamp
7. Student tool box

SAFETY

Be cautious when working around the hot water when purging the air from the air traps; it may burn you if it contacts the skin. Remember to wear safety goggles and protective gloves at all times.

PROCEDURE

1. Turn on the circulating pump(s).
2. Set the temperature control at the unit heaters low enough to circulate water through the units.
3. Power up the hot water boiler.
4. Install thermometers on supply water, return water, discharge and return air of the unit heaters.
5. Measure combustion efficiency.
6. Measure voltage and amp readings at boiler and circulating pump(s).
7. Purge all of the air from the system.
8. Record the following information:

Boiler Manufacturer _____________

Boiler model # _____________

Input BTU's _____________

Output BTU's _____________

Entering water temperature at the boiler ____________
Leaving water temperature at the boiler ____________
Discharge air temperature at unit heater ____________
Return air at unit heater ____________
Boiler amps ____________
Circulator pump amps ____________

9. Ask your instructor to install a problem symptom in the unit.
10. Test, repair, and record your conclusions.
11. Upon completion, remove all test instruments and equipment.

QUESTIONS

1. What type of hydronic system were you working on, series loop, one pipe, two pipe? Explain.

2. How efficient was the boiler combustion? How did you measure the efficiency?

3. What type of control was used to control temperature at the unit heaters?

4. Did both unit heaters produce equal amounts of heat?

5. What was the problem symptom, and what did you do to remedy the problem?

6. How does air get into a water system?

7. How can you get air out of a water system?

8. Why is an expansion tank with air in it necessary?

9. Why should you never allow cold water to enter a hot, empty hydronic system?

10. What is the purpose of a pressure relief valve?

Laboratory Worksheet - # A11

Student__________________________

Instructor __________________________

Date _____________ Grade__________

RESIDENTIAL CONTROL SYSTEMS

EXPECTED COMPETENCY:

To become acquainted with the basics of air conditioning electrical control systems

STUDY MATERIAL

Chapter H35

EQUIPMENT

1. Control panels containing the component parts of an air conditioning control system. Lights are used on the panels to designate electric motor loads.
2. Wiring diagrams of the panel
3. Control wire
4. Power wire
5. Power source of proper voltage, phase and frequency for the controls and loads used

SAFETY

Wear your safety goggles at all times, be cautious when working with the electrical loads to prevent short circuits or unintentional grounding.

PROCEDURE

1. Wire the power system of the panel according to the wiring diagram for the panel.
2. Wire the low voltage control system of the panel.
3. Before operating the panel, have the instructor check it out.
4. Operate the panel through the various steps of operation.
5. Repeat procedures 1 through 4 for each type of panel.

Laboratory Worksheet - # A12

Student_______________________

Instructor _____________________

Date ___________ Grade________

COMMERCIAL AND ENGINEERED CONTROL SYSTEMS

EXPECTED COMPETENCY:

To become acquainted with the basics of pneumatic control systems used in commercial and industrial fields

STUDY MATERIAL

Chapters A31 and H35

EQUIPMENT

1. Panel boards of pneumatic controls, systems and lights to simulate loads
2. Air pressure source
3. Supply of air pressure tubing and fittings
4. Supply of electrical control circuits

PROCEDURE

1. Connect the air system using plastic hose according to the piping diagram supplied for the panel.
2. Connect the electrical control circuits according to the diagram supplied with the panel.
3. Have the instructor check out the panel before operating.
4. Operate the panel through the on and off sequence of operation.
5. Record the step-by-step operating sequence.
6. Record the operating steps of each control and save for future reference .

QUESTIONS

1. What does the term "main air" refer to?

2. Pneumatic thermostats, relays and switches are called _______________.

3. Valve and motor loads are called____________ or ______________.

4. The air line between a thermostat and a valve is called a______________.

Laboratory Worksheet - # A13
Student__________________________
Instructor ________________________
Date ____________ Grade_________

HEATING, MEASURING, AND TESTING EQUIPMENT-GAS

EXPECTED COMPETENCY:
To become acquainted with gas heating test equipment and its use

STUDY MATERIAL
Chapters H36, H40, I47, and TR50

EQUIPMENT
1. Gas fired heating unit installation
2. Gas test kit
3. 12" rule
4. Clear plastic tube - 36"
5. Tape
6. Stop watch
7 Two pocket dial thermometers 0° to 200°F

SAFETY
Remember to wear to your safety goggles, and be cautious when working around the gas supply. Make all connections tight and leak-free.

PROCEDURE
1. Insert dial thermometers in supply and return air ducts in proper places.
2. Operate unit with thermostat set at maximum.
3. Use the natural gas meter to obtain proper input. Adjust regulator on the gas valve until proper input is obtained.
4. Turn gas valve to "pilot" setting and connect the gas pressure manometer.
5. Open the gas valve and measure manifold pressure. Record this pressure.
6. Shut off the gas valve, disconnect the gas pressure manometer. Connect the manometer made from the rule and plastic tube containing water.
7. Open the gas valve and measure the manifold pressure. Compare the readings of the two manometers.
8. Shut off the gas valve - remove the water manometer and fitting. BE SURE TO RE-INSTALL THE PLUG IN THE VALVE.
9. Open the gas valve and allow the unit to operate.

10. When the supply and return temperatures have stabilized:
    a. Measure the flue gas CO_2.
    b. Measure the stack temperature.
    c. Determine the efficiency of the unit.
11. Determine the temperature rise of the air through the unit.

12. If the temperature difference is not within data plate manufacturer's specifications, adjust the blower speed until a temperature difference as close as possible to design is obtained.
13. With the change in air quantity, re-run step 10.
14. Shut down unit - return the blower speed setting to original setting.

QUESTIONS

1. What is the allowable input range for a natural gas-fired heating unit?

2. With the proper input obtained by timing the meter, what was the operating manifold pressure?

3. What is the recommended stack temperature rise for proper operation of a standard gas-fired unit?

4. The stack temperature of a high efficiency gas furnace can be as low as __________°F.

5. What effect did a reduction in air over-the-unit have on the operating efficiency?

6. What is CO_2?

7. What is CO?

8. Which gas is considered highly dangerous, CO_2 or CO?

Laboratory Worksheet - # A14
Student________________________
Instructor _______________________
Date ____________ Grade_________

INSTALL AND OPERATE AN ATMOSPHERIC GAS BURNER WITH PILOT LIGHT

EXPECTED COMPETENCY:
To become acquainted with the steps necessary to connect, adjust and operate an atmospheric gas burner warm air furnace with standing pilot

STUDY MATERIAL
Chapters H36, H40, I47, and TR50

EQUIPMENT
1. Safety goggles
2. Gas-fired heating unit with standing pilot
3. Student tool set
4. Necessary supply pipe, new pipe and fittings
5. Stop watch
6. Two dial thermometers

SAFETY
Wear safety goggles at all times. Check for leaks before turning on the ignition.

PROCEDURE
1. Connect the fuel supply line to gas supply.
2. Check for leaks with soap suds or liquid leak detector - DO NOT USE OPEN FLAME.
3. Connect electrical supply to unit.
4. Connect thermostat to control system.
5. Connect the vent flue.
6. Purge the gas control system and light the pilot.
7. Time the gas supply meter and adjust the unit for proper input. Adjust for +/-10% to rated input. DO NOT OVER-FIRE THE UNIT.
8. Adjust burner primary air for proper fire.
9. Operate the unit and measure the temperature rise.
10. Adjust the blower speed for proper C.A.C. (Continuous air circulation) temperature rise.

QUESTIONS
1. The input on a natural gas unit is adjusted by the adjusting screw on the _____________.

2. The input on an L.P. gas unit is adjusted by the adjusting screw on the ______________.

3. The proper burner appearance on a vented type gas burning unit is _____________ .

4. The proper burner appearance on a non-vented gas burning appliance is ____________.

5. What is the proper size flame on a pilot burner?

6. The top of a vent must extend how far above the highest point of the roof?

7. When gas-fired units are used at high altitude, what is the rule for reducing the input?

8. When is it possible to take return air from the furnace room?

9. A gas-fired unit made up of several heat exchangers is called a ___________ type.

10. What is the purpose of the draft diverter?

11. What are the symptoms of a shortage of primary air?

12. Describe a gas flame that is starved for oxygen?

13. What does the draft diverter do in the case of a down draft in the flue pipe?

14. What do yellow tips mean on a gas flame?

15. What would cause too much temperature difference across the heat exchanger?

16. Can a pilot light flame be adjusted? If so, how?

17. What is the recommended manifold pressure for an LP gas furnace?

18. Which gas is heavier than air, natural or LP?

Laboratory Worksheet - # A15

Student__________________________

Instructor _________________________

Date ____________ Grade_________

INSTALL AND OPERATE AN ATMOSPHERIC GAS BURNER WITH ELECTRONIC PILOT

EXPECTED COMPETENCY:
To become acquainted with the steps necessary to connect, adjust and operate an atmospheric gas burning warm air furnace with electronic ignition

STUDY MATERIAL
Chapters H36, H40, I47, and TR50

EQUIPMENT
1. Safety goggles
2. Gas-fired heating unit with electronic ignition
3. Student tool set
4. Ncccssary supply pipe, vent pipe and fittings
5. Stop watch
6. Two dial thermometers

SAFETY
Wear safety goggles at all times. Check for leaks before turning on the ignition.

PROCEDURE
1. Connect the fuel supply line to the gas supply.
2. Check for leaks WITH SOAP SUDS OR LIQUID LEAK DETECTOR - DO NOT USE OPEN FLAME.
3. Connect electrical supply to unit.
4. Connect thermostat to control system.
5. Connect the vent flue.
6. Purge the gas supply system.
7. Turn on main power and lower thermostat setting to start unit.
8. If unit does not start, recycle unit until pilot is ignited and established.
9. Adjust the unit for proper input.
10. Adjust the burners for proper fire.
11. Operate the unit and measure temperature rise.
12. Adjust the blower speed for proper temperature rise.

QUESTIONS

1. On a gas-fired unit with electronic pilot, what is the proper main burner flame appearance?

2. Is it necessary to provide a separate ground wire in the electrical supply for electronic ignition units?

3. What is the proper percent of input for gas-fired units with electronic ignition?

4. For C.A.C. (continuous air circulation) the desired air temperature rise through the heating unit is ____________°F.

5. What is the output voltage of an electronic pilot ignition system?

Laboratory Worksheet - # A16

Student____________________________

Instructor ____________________________

Date ______________ Grade__________

HEATING SERVICE AND TROUBLE ANALYSIS-GAS

EXPECTED COMPETENCY:

To become acquainted with the steps necessary to solve problems in gas-fired heating systems

---

STUDY MATERIAL

Chapters H36, H40, I47, and TR50

EQUIPMENT

1. Gas-fired heating unit which has been instructor-rigged to simulate problems
2. Gas-fired test kit
3. Two dial thermometers
4. Student tool kit
5. Stop watch
6. Test sheets (copies from text book in assigned chapters)

SAFETY

Wear safety goggles at all times and be sure to inspect all gas lines for leaks prior to ignition using soap bubbles or electronic leak detectors (designed to locate natural gas leaks).

PROCEDURE

1. Check and test the unit to obtain proper operation. Use the procedures outlined in text book.
2. Repeat the procedure for each problem set up by the instructor.

Laboratory Worksheet - # A17
Student___________________________
Instructor _______________________
Date ____________ Grade_________

HEATING, MEASURING, AND TESTING EQUIPMENT-OIL

EXPECTED COMPETENCY:

To become acquainted with oil heating, testing equipment and its use

STUDY MATERIAL

Chapters H36, H40, I47, and TR50

EQUIPMENT

1. Oil-fired heating unit installation
2. Oil test kit
3. Fuel unit pressure kit
4. Nozzle gauge or steel rule
5. Nozzle wrench
6. Two dial thermometers
7. Student tool kit
8. Flame mirror
9. Fuel oil cap

SAFETY

Wear safety goggles at all times and clean up any oil before and after the exercise.

PROCEDURE

1. Remove the firing head from the oil burner, check for nozzle. Some units are shipped without a nozzle.
2. Using the nozzle wrench, remove the nozzle, remove the input screen and inspect. DO NOT REMOVE THE SWIRL STEM. IF SWIRL STEM BECOMES PLUGGED, REPLACE THE NOZZLE ASSEMBLY. Make sure the nozzle is the correct size, spray angle and cone type for the burner.
3. Re-assemble this nozzle and put back in place on the firing head.
4. Measure and position the ignition electrodes.
5. Re-install the firing head into the burner.
6. Connect the oil pressure gauge.
7. Loosen the fuel unit purge plug and operate until the fuel unit is properly purged.
8. Operate the unit and set nozzle pressure at proper setting.

9. When the supply and return air temperatures have stabilized:
   a. Check CO_2.
   b. Check stack temperature rise.
   c. Check over-fire draft.
   d. Check stack draft.
   e. Determine the operating efficiency of the unit.
   f. Take a smoke test and record.
10. Set the unit for the proper:
    a. CO_2
    b. Over-fire draft.
11. After temperatures have stabilized, re-run step 9. Record the difference in operating efficiency and air temperature rise.
12. Set the air temperature rise to obtain as close to C.A.C. temperature rise as possible.
13. Re-run step 9. Record the difference in operating efficiency and smoke test.
14. Shut down the unit. Return all the burner and blower speed settings to original.

QUESTIONS

1. What is the normal nozzle operating oil pressure?

2. With a round combustion chamber, what nozzle spray angle would normally be used?

3. With a long narrow rectangular combustion chamber, what nozzle spray angle would normally be used?

4. What is the maximum allowable difference between the over-fire draft and the stack draft?

5. What is the normal stack temperature rise of a standard oil-fired unit?

6. In residential oil-fired units, the smoke test should indicate a Number_________ smoke.

Laboratory Worksheet - # A18
Student____________________________
Instructor __________________________
Date ____________ Grade__________

INSTALL AND OPERATE AN ATOMIZING TYPE OIL-FIRED HEATING UNIT

EXPECTED COMPETENCY:
To become acquainted with the steps necessary to connect, adjust and operate an atomizing type oil-fired heating unit

STUDY MATERIAL
Chapters H36, H40, I47, and TR50

EQUIPMENT
1. Atomizing oil furnace and oil piping supplies, as required
2. Electrical connections

SAFETY
Wear your safety goggles at all times, check all oil fittings for leaks, and clean up any excess oil before and after the exercise. Have your instructor check all fitting and electrical connections prior to start up.

PROCEDURE
1. Connect the oil supply line to the oil burner fuel unit.
2. Connect the barometric control to the flue outlet of the unit.
3. Connect the flue vent to the barometric control.
4. Connect 120-watt power to the heating unit.
5. Connect a heating thermostat to the Protecto-relay control.
6. Connect a pressure gauge to the fuel unit.
7. Remove the firing head from the oil burner and check for:
   a. Proper nozzle in place.
   b. Proper ignition electrode placement.
8. Re-install the firing head.
9. Start and purge the fuel unit.
10. Operate the unit to establish fire and adjust for proper oil pressure and minimum observed smoke through observation port.
11. Run efficiency test on the unit and set for proper CO2, over-fire draft and stack temperature.
12. Determine the operating efficiency of the unit.

QUESTIONS

1. Most oil-fired units operate with what "over-fire" draft pressure?
2. The normal nozzle pressure on an atomizing burner is _______PSIG.
3. The desired operating CO2 range of a residential oil-fired unit is ________ to _______%.
4. The normal stack temperature rise is from ________to _______°F.

Laboratory Worksheet - # A19
Student______________________
Instructor ______________________
Date ____________ Grade_________

HEATING SERVICE AND TROUBLE ANALYSIS-OIL

EXPECTED COMPETENCY:
To become acquainted with the steps necessary to solve problems in oil-fired heating systems

STUDY MATERIAL
Chapters H36, H40, I47, and TR50

EQUIPMENT
1. Oil-fired heating unit which has been instructor-rigged to simulate problems
2. Oil-fired test kit
3. Two dial thermometers
4. Student tool kit
5. Oil pressure test kit
6. Test sheets (copies from text book)

SAFETY
Be cautious when troubleshooting the equipment. Keep safety goggles on at all times. Always use insulated tools and meter leads.

PROCEDURE
1. Check and test the unit to obtain proper operation. Use the procedure outlined in text book.
2. Repeat the procedure for each problem set up by the instructor.

Laboratory Worksheet - # A20

Student_________________________

Instructor ________________________

Date ____________ Grade_________

ADJUSTING AIR CONDITIONING SYSTEMS FOR PROPER COOLING LOAD

EXPECTED COMPETENCY:

To become acquainted with the method of determining and adjusting for the proper air quantity through the DX oil

STUDY MATERIAL

Chapters A30, A32, H40, I47, TR49

EQUIPMENT

1. Operating air conditioning system
2. Four dial thermometers
3. Sling psychrometer

PROCEDURE

1. Use the sling psychrometer to obtain the conditions of the air in the conditioned area. This is the cooling load to be handled.
2. Insert thermometers in the following locations:
   a. Return air grille
   b. Return air duct at the coil
   c. Supply air duct at the coil
   d. Supply air grille
3. When all thermometers have reached their ultimate reading, determine the temperature drop of the air through the coil. (Return air temperature minus supply air temperature equals temperature difference of the coil.)
4. Using the chart in text, determine the proper temperature drop required for the conditions of the air in the conditioned area.
5. Adjust the blower speed up or down to obtain as close a temperature drop as possible to the desired amount.
6. Compare the return air temperature to the temperature of the air entering the coil. If air temperature rise is excessive, check for air leaks in the duct or lack of insulation.
7. Compare the supply air temperature to the temperature of the air leaving the coil. If temperature rise is excessive, check for lack of duct insulation.

QUESTIONS

1. What is the purpose of establishing the proper temperature difference of the air through the DX coil?

2. When is this air temperature drop test necessary?

3. What is the maximum air temperature rise allowable in the return air duct?

4. What is the maximum air temperature rise allowable in the supply air duct?

5. At 80° D.B. and 50% R.H., a 5° F rise in the supply air temperature would cause what percent of capacity loss of the unit?

6. If the return air duct and leakage raised the coil air temperature 5°F over the conditioned air temperature, what percent of load increase would the unit have to be able to handle?

Laboratory Worksheet - # A21
Student______________________
Instructor ______________________
Date ___________ Grade_________

DETERMINE SYSTEM AIR QUANTITY BY AIR DUCT FLOW RATES

EXPECTED COMPETENCY:

To become acquainted with the instruments and method used to measure air quantities in duct work

STUDY MATERIAL

Chapters H40 and I47

EQUIPMENT

1. Operating air conditioning system
2. Inclined manometer
3. Pitot tube set
4. Steel measuring ruler
5. Anemometer
6. Stop watch
7. Flow hood air quantity meter

PROCEDURE

1. Adjust the air quantity through the DX coil for the conditions of the air in the conditioned area.
2. Determine the air flow through the supply air and return air ducts via the Pitot tube method.
   a. Measure the air velocity through the duct in feet per minute.
   b. Determine the square foot area of the duct.
   c. Determine the cubic feet per minute of the airflow through the duct.
3. Determine the total quantity of air through the supply duct system by use of the anemometer and stop watch.
   a. Measure the quantity of air from each supply outlet for one minute. (Use the total quantity for a five-minute period on each outlet and divide by five minutes to obtain the average per minute).
   b. Total the quantities per outlet to obtain system total.
4. Repeat the procedure for each return air grille.
5. If more than 5% difference on air quantity between supply and return, check for air leaks in the duct system.
6. Substitute the air flow hood for the anemometer and stop watch and repeat steps 3, 4 and 5.

QUESTIONS

1. Which method of determining the total air quantity is the most accurate?

2. When is it necessary to know the actual quantity of air flowing through the system?

Laboratory Worksheet - # A22
Student___________________________
Instructor ___________________________
Date ______________ Grade___________

START UP AND CHECK OUT AN AIR-COOLED AIR CONDITIONING UNIT

EXPECTED COMPETENCY:
To become acquainted with the steps necessary to check and adjust an air conditioning unit for peak efficiency (THIS EXPERIMENT SHOULD BE RUN ON VARIOUS TYPE UNITS.)

STUDY MATERIAL
Chapters I47 and TR49

EQUIPMENT
1. Air-cooled air conditioning system with air distribution duct system
2. Pressure gauge/manifold set
3. Sling psychrometer
4. Seven dial pocket thermometers
5. Clamp type volt/ammeter
6. Student tool kit
7. Test sheets (copies of test sheet information from text book)

PROCEDURE
1. Mount dial thermometers on air discharge opening of condensing unit.
2. Tape thermometer to liquid line at outlet of condenser.
3. Install high pressure and compound gauges.
4. Clamp ammeter, set on highest range, around one of the power leads to the condensing unit.
5. Install thermometers in supply and return ducts.
6. Use the test sheet duplicates of the test information from text book; record the unit information.
7. Start the unit using, steps 9 and 10.
8. Adjust the system for proper air quantity.
9. Determine the DX coil operating super heat.
10. Determine the condenser split.
11. Determine the liquid sub-cooling.
12. Calculate the condenser air discharge temperature and condenser air temperature rise per the text book.
13. Calculate the net capacity of the system.
14. Compare the net capacity to the manufacturer's literature.

QUESTIONS

1. Does the capacity of the unit come within +/- 10% of the manufacturer's literature? If not, why not?

2. What should be the range of the coil super heat?

3. What should be the range of liquid sub-cooling?

4. What is the normal condenser split range for standard units?

5. What is the normal condenser split range for high efficiency units?

Laboratory Worksheet - # A23

Student_______________________

Instructor _______________________

Date ____________ Grade_________

TEST SHEET
HEATING - GAS

EXPECTED COMPETENCY:

To understand the requirements to operate and test a gas-fired furnace

1. Unit Model No. __________S/N
2. Type __________________ Input ______________B.T.U.H.
3. Type of gas: Nat._____ Mixed_____ Mfg_____ Propane_____ Butane _____
4. Heat Content - B.T.U. per cubic foot _________
5. Specific gravity of the gas _________
6. Main burner orifice drill size: Found___________ Left _________
7. Manifold pressure "W.C.": Found ___________ Left _________
8. Meter test dial size: _________cubic feet per revolution.
9. Seconds required per 1 R.P.M.: Found ________ Left _______
10. Flame before adjustment: Sharp blue ________ Soft blue ________
    Yellow tips ________ after adjustment: ________ Soft blue________
11. Neutral point adjustment:
    a. Factory designed, not adjustable ____________
    b. Conversion burner: (1) Found: Below adjustment range ___________
       Above adjustment range___________
       In correct adjustment range__________

12. Air temperature rise

	1st test	2nd test	3rd test
Supply air temperature	______°F	______°F	______°F
Return air temperature	______°F	______°F	______°F
Temperature rise	______°F	______°F	______°F

13. CO_2: 1st test ________%; 2nd test _______%; Left _______ %

14. Stack temperature rise

	1st test	2nd test	3rd test
Stack temperature	______°F	______°F	______°F
Combustion air temperature	______°F	______°F	______°F
Temperature rise	______°F	______°F	______°F

15. Combustion efficiency

	1st test	2nd test	3rd test
% of efficiency	______%	______%	______%

Laboratory Worksheet - # A24
Student__________________________
Instructor ________________________
Date ____________ Grade_________

TEST SHEET
HEATING - OIL

EXPECTED COMPETENCY:
To understand the requirements to operate and test an oil-fired furnace

1. Make of burner: ________________

2. Model no. ____________________ S/N _____________________

3. Nozzle: Size _________ G.P.H. type ___________ Spray angle ___________.

4. Refractory: Material _____________ Shape _______________

5. Over-fire draft: ___________" W.C.

6. Stack draft: _____________" W.C.

7. Heat exchanger flow resistance:____________ " W.C.

8. CO_2: __________ %

9. Stack temperature rise:
   a. Stack temperature: _________°F
   b. Combustion air temperature: _________°F
   c. Net stack temperature: _________°F

10. Burner efficiency: __________ %

11. Smoke number:____________

12. Air temperature rise:
    a. Supply air temperature:________°F
    b. Return air temperature: ________°F
    c. Air temperature rise: __________°F

Laboratory Worksheet - # A25
Student_________________________
Instructor _______________________
Date ____________ Grade_________

TEST SHEET
AIR CONDITIONING

EXPECTED COMPETENCY:
To understand the requirements to operate and test an air conditioner

1. Outdoor unit: Model no. _________ S/N ___________

2. Installation date:____________________

3. Type of pressure reducing device:
   a. TX valve: ____________
   b. Capillary tube: ____________

4. Type of fuse protection: ____________capacity __________ amps

5. Wire size to outdoor unit: _______________

6. Condition of air filter: __________________ New ________________

7. Voltage at outdoor unit: Unit off ____________volts

8. Voltage at unit: Start up ______________volts

9. Amperage draw of unit at:
   a. Start up (L.R.A.) ________amperes
   b. Running (F.L.A.)________amperes

10. Return air grille temperature: __________°F

11. Return air grille relative humidity:_________%

12. Return air temperature at coil: ____________°F

13. Supply air temperature off coil: ___________°F

14. Temperature drop of air through coil: _________°F

15. Air temperature at supply grille: ___________°F

16. Suction line temperature at coil: ___________°F

17. Comp. suction pressure (DX coil evaporating temperature)
_____________P.S.I.G. ___________ °F

18. DX coil super heat: ___________

19. Temperature of air entering condenser: _________°F

20. Condensing pressure __________P.S.I.G. Condensing temp.____________°F

21. Condenser split:_________________°F

22. Liquid line temperature: ___________°F

23. Liquid sub-cooling: ___________°F

24. Condenser discharge air temperature: ___________°F

25. C.F.M. through condensing unit:________________

26. Unit gross capacity:____________B.T.U.H.

27. Motor heat input: ______________B.T.U.H.

28. Unit net capacity: ______________B.T.U.H.

Laboratory Worksheet - # HP1
Student____________________________
Instructor __________________________
Date _____________ Grade__________

BASIC CONSTRUCTION OF HEAT PUMPS

EXPECTED COMPETENCY:
To become acquainted with the construction and parts arrangement of various types of heat pumps

---

STUDY MATERIAL
Chapter HP42.

EQUIPMENT

1. Air to air split heat pump system
2. Air to air package heat pump unit
3. Liquid to air split heat pump system
4. Liquid to air package heat pump unit
5. Liquid to liquid package heat pump unit

SAFETY
Be cautious when working with refrigerant pressures, especially the discharge lines. These pressures can reach up to 300 PSIG during high ambient conditions. Use quick-release hose adaptors to minimize refrigerant loss and avoid liquid refrigerant burns. Always wear safety goggles.

PROCEDURE
Inspecting each type unit, record the following information:

1. Location of the compressor
2. Location of the reversing valve
3. Type of pressure reducing device on the:
   a. Inside coil
   b. Outside coil
4. Construction of the inside coil
5. Construction of the outside coil
6. Type of check valve on the:
   a. Inside coil
   b. Outside coil
7. Starting at the compressor discharge, trace the refrigerant flow (listing on a separate piece of paper) through the system on the heating cycle.
8. Repeat step 7 on the cooling cycle.
9. Repeat steps 7 and 8 for each type of unit.

QUESTIONS

1. Which check valve is open on the heating cycle?

2. Which check valve is open on the cooling cycle?

3. On the heating cycle, the hot discharge gas from the compressor goes to which coil?

4. Which direction does the refrigerant vapor travel through the compressor on the heating cycle?

5. Which direction does the refrigerant vapor travel through the compressor on the cooling cycle?

6. On the split heat pump system, is the large refrigerant line the suction or hot gas line on the heating cycle?

7. Refrigerant pressure in this line on the heating cycle would be close to what pressure - suction or discharge?

Laboratory Worksheet - # HP2
Student________________________
Instructor ______________________
Date ____________ Grade_________

INSTALL AND CHECK OUT A HEAT PUMP THERMOSTAT

EXPECTED COMPETENCY:

To become acquainted with the wiring and operating circuits of a heat pump thermostat

STUDY MATERIAL

Chapter HP44

EQUIPMENT

1. Multi-stage heat pump thermostat and sub-base
2. Heat pump manufacturer's wiring diagram
3. Air to air split heat pump system

PROCEDURE

1. Following the manufacturer's wiring diagram, conncct thc thermostat to the air handler and to the outside unit. Wire one circuit at a time.
2. Tracc thc wiring, following the wiring diagram.
3. Have the installation inspected by the instructor before operating.
4. Operate the thermostat through all the operating steps on both heating and cooling cycles.

QUESTIONS

1. How many degrees of temperature setting are between the heating and cooling settings? Why?

2. If the emergency heat switch is set to the "Emergency" setting, heat is supplied by what part of the system?

3. What is the difference in temperature maintained by the 1st stage and 2nd stage of heat?

4. The first stage of cooling controls what part of the system and performs what function?

5. What is the advantage of a heat pump on the cooling cycle as compared to an air conditioning system?

Laboratory Worksheet - # HP3

Student________________________

Instructor _______________________

Date ____________ Grade_________

HEAT PUMP DEFROST CONTROLS

EXPECTED COMPETENCY:

To become acquainted with the parts and operation of defrost control systems

STUDY MATERIAL

Chapter HP44

EQUIPMENT

1. Samples of defrost systems:
   a. Temperature initiation - temperature termination
   b. Pressure initiation - temperature termination
   c. Time initiation - temperature termination
   d. Pressure - time initiation - temperature termination

SAFETY

Be cautious when working with low-voltage solid state defrost controls. To avoid damaging the electronic circuits, turn off power prior to your inspection.

PROCEDURE

1. Examine each control system and record the part number and its function.

Part name	Part no.	Function
________________	__________	_____________
________________	__________	_____________
________________	__________	_____________
________________	__________	_____________

2. Repeat this procedure until all parts of all control systems have been recorded. (Keep these sheets for future reference.)

QUESTIONS

1. What function is common to all heat pump systems?

2. The successful operation of the temperature initiation - temperature termination type defrost control depends heavily upon the condition of the air filter in the air handler. True or False?

3. Wind direction and speed can have an adverse effect on which of the four types of defrost control systems?

4. Which defrost control system is prone to give false or unnecessary defrost cycles?

5. What is the purpose of the auxiliary defrost relay addition to the defrost control system?

Laboratory Worksheet - # HP4
Student_______________________
Instructor ____________________
Date ____________ Grade_________

START UP, CHECK OUT, AND OPERATE AN AIR TO AIR HEAT PUMP-COOLING CYCLE (Outside air above 65 °F)

EXPECTED COMPETENCY:
To become acquainted with the steps necessary to start-up, checkout and determine the operating capacity of an air to air split system - heat pump system on the cooling cycle

STUDY MATERIAL
Chapters HP44, I47 and TR51

EQUIPMENT
1. Air to air split system heat pump with connected air distribution system
2. Student tool kit
3. Sling psychrometer
4. Heat pump style pressure gauge set
5. Clamp type ammeter
6. Voltmeter
7. Manufacturer's literature on the unit
8. Seven dial-type thermometers or electronic thermometers

PROCEDURE
1. Connect the gauge set to the unit.
2. Set the ammeter to the highest amperage scale and clamp to one of the hot power leads to the compressor contactor.
3. Install the thermometers:
   a. Three on the outdoor coil inlet. They must be in proper location.
   b. One thermometer on the liquid line at the liquid outlet of each coil.
   c. One thermometer in each of the return air and supply air ducts to the inside air handler.
4. Before starting the unit, record all the information on the test sheet, items 1 through 10.
5. Start the unit on the cooling cycle. WATCH THE AMMETER TO MAKE SURE COMPRESSOR MOTOR STARTS PROPERLY.
6. After the unit has balanced out, record the information on the test sheet-items 11 through 27.
7. The check-out information for C.F.M. of air through the inside coil is different for cooling than for heating. Cooling - adjust the air quantity through the coil for the desired temperature difference for the return air conditions the same as for a standard air conditioning system.

8. When the temperature difference of the air through the inside coil has been established, measure and calculate the liquid sub-cooling. Adjust the refrigerant charge to bring the sub-cooling within the proper range.
9. Determine the net operating capacity of the system.
10. Capacity results should come within +/- 10% of manufacturer's rating.

QUESTIONS

1. With the liquid sub-cooling and the temperature difference of the air through the inside coil in the normal range, what is the operating super heat of the suction gas off the inside coil?

2. With the same conditions as in question 1, what is the net capacity of the system?

3. What is the "split" of the outside coil?

4. The large vapor line pressure is closest to which pressure, suction or discharge?

5. In this particular unit, to operate on the cooling cycle, is the reversing valve energized or de-energized?

Laboratory Worksheet - # HP5
Student___________________________
Instructor ________________________
Date ____________ Grade__________

START UP, CHECK OUT, AND OPERATE AN AIR TO AIR HEAT PUMP-HEATING CYCLE (Outside air below 65 °F)

EXPECTED COMPETENCY:
To become acquainted with the steps necessary to start up, check out and determine the operating capacity of an air to air split system, heat pump system on the heating cycle

STUDY MATERIAL
Chapters HP44, I47 and TR51

EQUIPMENT
1. Air to air split system, heat pump system with connected air distribution system
2. Student tool kit
3. Heat pump style pressure gauge set
4. Clamp type ammeter
5. Voltmeter
6. Four dial-type thermometers
7. Manufacturer's literature on the unit

SAFETY
Remember to wear your safety goggles, and be careful when handling refrigerant hoses to prevent leaking refrigerant to the atmosphere.

PROCEDURE
1. Connect the gauge set to the unit.
2. Set the ammeter to the highest amperage scale and clamp it to one of the hot power leads to the compressor contactor.
3. Install the thermometers:
   a. Three on the outdoor coil inlet. They must be in proper location.
   b. One thermometer on the liquid line at the liquid outlet of each coil.
   c. One thermometer in each of the return air and supply air ducts to the inside air handler.

Before starting the unit, record all the information on the test sheet, items 1 through 10.

4. Start the unit on the heating cycle. WATCH THE AMMETER TO MAKE SURE THE COMPRESSOR MOTOR STARTS PROPERLY.
5. After the unit has balanced out, record the information on the test sheet-items 11 through 27.

6. The check-out information for C.F.M. of air through the inside coil is different for heating than for cooling.
   a. Install dial thermometers in the supply and return air ducts at the inlet air handler.
   b. With the unit operating in the emergency heat mode (using the auxiliary electric heat elements) measure the voltage and amperage to the auxiliary heat unit.
   c. After the air temperatures have stabilized, determine the C.F.M. through the coil using the formula outlined in the text book. Record this amount on the test sheet.
7. Calculate the gross B.T.U.H. capacity of the unit using the formulae in the text.
8. Compare the results against the manufacturer's rating for the outdoor unit ambient temperature measured during the test. The capacity of the unit should be within +/- 10% of the manufacturer's rating.

QUESTIONS

1. What is the amount of liquid sub-cooling compared to the sub-cooling on the cooling cycle?

2. What is the split of the condenser on the heating cycle?

3. The large vapor line pressure is closest to which pressure - suction or discharge?

4. What is the operating super-heat of the evaporator on the heating cycle?

Laboratory Worksheet - # HP6

Student______________________

Instructor ____________________

Date ___________ Grade_________

START UP, CHECK OUT, AND OPERATE A LIQUID TO AIR HEAT PUMP

EXPECTED COMPETENCY:

To become acquainted with the steps necessary to start up, check out and determine the operating capacity of a liquid to air heat pump on both heating and cooling cycles

STUDY MATERIAL

Chapters HP44, I47 and TR51

EQUIPMENT

1. Operating liquid to air heat pump system
2. Student tool kit
3. Heat pump style pressure gauge set
4. Clamp type ammeter
5. Voltmeter
6. Four dial type thermometers
7. Stop watch
8. Liquid bucket of known capacity
9. Sling psychrometer
10. Manufacturer's literature on the particular unit

SAFETY

Be cautious when working with the clamp-on ammeter to only test a single wire at a time, and to only touch the insulated portions of the wire.

PROCEDURE

1. Connect the gauge set to the unit.
2. Set the ammeter to the highest amperage scale and clamp to one of the hot power leads to the compressor contactor.
3. Install the thermometers:
   a. One thermometer strapped to the liquid supply and return lines to the tube-in-a-tube heat exchanger in the unit.
   b. One thermometer in each of the return air and supply air ducts from the inside coil.
   c. One thermometer at the outlet of the inside coil.
4. Before starting the unit, record all the information on the test sheet, items 1 through 10.
5. Set the thermostat to operate on the cooling cycle.

6. Start the unit, observing the action of the ammeter TO MAKE SURE THE COMPRESSOR MOTOR STARTS PROPERLY.
7. When the supply and return air thermometers have stabilized, determine the temperature drop of the air through the inside coil.
8. Determine the required temperature drop to be obtained to satisfy the load in the conditioned area. Adjust the C.F.M. through the coil to obtain as close a temperature drop as possible.
9. Set the liquid supply valve to maintain 105°F condensing temperature.
10. Record the information in items 11 through 29 of the test sheet-cooling cycle.
11. Determine the net capacity of the unit as outlined in text.
12. Set the thermostat to operate on the heating cycle.
13. Adjust the liquid supply regulating valve to maintain a 45° evaporator.
14. Complete the items 11 through 29 on the test sheet - heating cycle.
15. Determine the unit set capacity by the method outlined in text book.

Laboratory Worksheet - # HP7

Student________________________

Instructor ______________________

Date ___________ Grade_________

TEST SHEET
HEAT PUMP - AIR TO AIR
COOLING

Outdoor unit: 1. Model no. ___________ 2. S/N ____________

Indoor unit: 3. Model no. ___________ 4. S/N ____________

5. Type of pressure reducing device:
   a. Outdoor unit: TX valve ___________ Capillary tube __________
   b. Indoor unit: TX valve ____________ Capillary tube __________

6. Type of overload protection: _______ Size _______ Amperes

Wire size:

7. To outdoor unit _______ 8. To indoor unit _______

9. Condition of air filter _______

Line voltage:

10. Unit off _______volts 11. Unit start _______ volts

12. Amperage draw: Start _______ amperes Run _______ amperes

13. Return air at return air grille:
    a. Temperature _______°F
    b. Relative humidity _______%

14. Return air at inside coil: _______°F

15. Supply air temperature at inside coil: _______ °F

16. Supply air temperature at supply grille: _______°F

17. Suction line temperature - inside coil: _______°F

18. Compressor suction pressure: _______ P.S.I.G.

19. Inside coil operating temperature: ________°F

20. Inside coil operating super heat: _________°F

21. Suction line temperature at compressor: ________°F

22. Temperature of air entering outside coil: ________°F

23. Compressor discharge pressure: ________ P.S.I.G.

24. Condensing temperature: Outside coil ________°F

25. Outside coil - operating temperature split: ________°F

26. Liquid sub-cooling: ________°F

27. CFM through outside unit ________

28. Unit gross capacity: ________ B.T.U.H.

29. Motor heat input: ________ B.T.U.H.

30. Unit net cooling capacity: ________ B.T.U.H.

Laboratory Worksheet - # HP8
Student___________________________
Instructor _________________________
Date ____________ Grade_________

TEST SHEET
HEAT PUMP - AIR TO AIR
HEATING

Outdoor unit: 1. Model no. ____________ 2. S/N _____________
Indoor unit: 3. Model no. ____________ 4. S/N _____________

5. Type of pressure reducing device:
   a. Outdoor unit: TX valve ____________ Capillary tube ___________
   b. Indoor unit: TX valve _____________ Capillary tube ___________

6. Type of overload protection: ________ Size ________ amperes

Wire size:
7. To outdoor unit ________ 8. To indoor unit ________

9. Condition of air filter ________

Line voltage:
10. Unit off ________volts 11. Unit start ________ volts

12. Amperage draw: Start ________ amperes Run ________ amperes

13. Air temperature at return air grille: ___________°F

14. Return air temperature at inside coil:___________ °F

15. Supply air temperature off inside coil:___________°F

16. Supply air temperature at supply grille:___________°F

17. Suction line temperature at outside coil:___________°F

18. Compressor suction pressure ___________ P.S.I.G.

19. Refrigerant boiling point ___________°F

20. Outside coil super heat ___________°F

21. Suction line temperature at compressor___________°F

22. Temperature air entering outside coil___________°F

23. Compressor discharge pressure ___________ P.S.I.G.

24. Condensing temperature inside coil ___________°F

25. Inside coil ambient temperature ___________°F

26. Liquid refrigerant sub-cooling ___________°F

27. Inside coil discharge air temperature_________

28. C.F.M. through inside coil: ______________C.F.M.

a. Electric auxiliary heat method:
   1. Aux heat:______volts, ______amps
   2. Air temp rise: __________
   3. Calculated C.F.M.________

b. Fossil fuel heat method
   1. Input to heating unit: ___________ B.T.U.H.
   2. Operating efficiency of heating unit:___________%
   3. B.T.U.H. output of heating unit:___________
   4. Air temperature rise ___________
   5. Calculated C.F.M. ___________

c. Static pressure method
   1. Supply air pressure:___________ "W.C.
   2. Return air pressure ___________ "W.C.
   3. Total external static pressure: ___________ "W.C.
   4. C.F.M. from manufacturer's literature: _________C.F.M.

29. Gross heating capacity: _____________ B.T.U.H.

Laboratory Worksheet - # HP9
Student_______________________
Instructor _____________________
Date ___________ Grade_________

TEST SHEET
HEAT PUMP - LIQUID TO AIR
COOLING

Outdoor unit: 1. Model no. ___________ 2. S/N ____________
Indoor unit: 3. Model no. ___________ 4. S/N ____________

5. Type of pressure reducing device:
   a. Outdoor unit: TX valve ___________ Capillary tube ___________
   b. Indoor unit: TX valve ____________ Capillary tube ___________

6. Type of overload protection: _______ Size _______ amperes

Wire size:
7. To outdoor unit _______ 8. To indoor unit _______

9. Condition of air filter _______

Line voltage:
10. Unit off _______volts 11. Unit start _______ volts

12. Amperage draw: Start _______ amperes Run _______ amperes

13. Air temperature at return air grille: __________°F

14. Return air temperature at inside coil:__________ °F

15. Supply air temperature off inside coil:__________°F

16. Supply air temperature at supply grille:__________°F

17. Suction line temperature at outside coil:__________°F

18. Compressor suction pressure __________ P.S.I.G.

19. Inside coil operating temperature _________ °F

20. Inside coil operating super heat _________°F

21. Suction line temperature at compressor __________ °F

22. Temperature of liquid entering outside coil __________ °F

23. Compressor discharge pressure __________ PSIG

24. Condensing temperature - outside coil __________ °F

25. Outside coil - operating temperature split __________ °F

26. Refrigerant liquid sub-cooling __________ °F

27. Temperature of liquid leaving outside coil __________ °F

28. Weight of liquid through outside coil __________ #/hr

29. Specific heat of liquid through outside coil __________ B.T.U./#

30. Unit gross capacity __________ B.T.U.H.

31. Motor heat input __________ B.T.U.H.

32. Unit net cooling capacity __________ B.T.U.H.

Laboratory Worksheet - # HP10
Student_______________________________
Instructor _____________________________
Date _____________ Grade__________

TEST SHEET
HEAT PUMP - LIQUID TO AIR
HEATING

Outdoor unit: 1. Model no. ____________ 2. S/N _____________
Indoor unit: 3. Model no. ____________ 4. S/N _____________

5. Type of pressure reducing device:
   a. Outdoor unit: TX valve ____________ Capillary tube ___________
   b. Indoor Unit: TX valve _____________ Capillary tube ___________

6. Type of overload protection: ________ Size ________ amperes

Wire size:
7. To outdoor unit ________ 8. To indoor unit ________

9. Condition of air filter ________

Line voltage:
10. Unit off ________volts 11. Unit start ________ volts

12. Amperage draw: Start ________ amperes Run ________ amperes

13. Air temperature at return air grille: ___________°F

14. Return air temperature at inside coil:___________ °F

15. Supply air temperature off inside coil:___________°F

16. Supply air temperature at supply grille: ___________°F

17. Suction line temperature - outside coil ___________ F

18. Compressor suction pressure ___________ PSIG

19. Outside coil operating temperature ___________°F

20. Outside coil operating super heat ___________°F

21. Suction line temperature at compressor ___________°F

22. Temperature of liquid entering outside coil ___________°F

23. Compressor discharge pressure ___________ PSIG

24. Condensing temperature - inside coil ___________°F

25. Operating temperature split - inside coil ___________ °F

26. Refrigerant liquid sub-cooling ___________ °F

27. Temperature of liquid leaving outside coil ___________ °F

28. Weight of liquid through outside coil ___________#/hr

29. Specific heat of liquid through outside coil ___________ B.T.U./#

30. Unit net capacity ___________ B.T.U.H.

31. Motor heat input ___________ B.T.U.H.

32. Unit gross capacity ___________ B.T.U.H.

33. C.F.M. through inside coil ___________ C.F.M.